Planes, Trains,
Automobiles, & More!

Planes, Trains, Automobiles, & More!

Frederick J. Hooven and His Brilliant Inventions

Becca Braun

Braun Ink
Cleveland

To the engineers, inventors,
and innovators of Ohio

Braun Ink is the publisher of the book. Please send all requests to reproduce information from this book and for bulk sales and special sales to:

Braun Ink
2618 N. Moreland Blvd
Cleveland, Ohio 44120
info@braunink

ISBN: 978-1-7355999-2-2

Book design by Ben Small, Live Publishing Co. Cover art created by Ben Small. Cover images come from photos section of this book. Permissions info. available in that section.

✿

Contents

Foreword

I started thinking of writing a book about my grandad, Fred Hooven, after a conversation I had with a friend who had served with me on various Ohio-based commissions and boards for over 25 years. We were donating our time to the organizations to help build the infrastructure in Ohio for innovation and business formation and growth. I was talking about Grandad's inventions and his roots in Dayton with the Wright brothers, when my friend asked, "Why don't you write a book about him?" We talked about how satisfying it would be to let people know about someone who was a great inventor and innovator and who had roots in Ohio. I also felt it would be wonderful for my friends and family to learn more about the career of a remarkable man, whom we knew as Fred, Dad, or Grandad.

My friend recommended Rebecca Braun, the author of this book, and she and I set out to put together a chronicle of Grandad's life through interviews, documents, and pictures. I've had a lot of fun reconnecting with some of Grandad's friends and protégés and, of course, my extended family members. Because Grandad was such a humble person, he never talked to me about any of his inventions. I've learned a lot about his career and additional inventions I was not aware of, and I've learned more about our family.

My grandad has always been one of my heroes, and his constant desire to learn about the world and to create things that make it better have inspired me throughout my life. I hope you will feel a bit of that inspiration as well.

Introduction

F red Hooven invented, designed, developed, and produced products across an array of fields such as automotive, aviation, biomedical, printing, acoustics, and environmental. The depth of his work extended beyond the realm of work-a-day or incremental improvements on devices and processes into the realm of meaningful, breakthrough inventions. He also played the chief-cook-and-bottle-washer role with his many inventions. Beyond being a signatory to a patent filing, he was an impassioned creator of the device or system at hand. He was a mechanical and electrical engineer, which meant that he had the ability to fully design every aspect of his inventions, and he was an extraordinary illustrator, musician, and writer, able to articulate and show the working power of his inventions. He had the holistic ability to conceive of and build a prototype, revise it, build a final device, and develop a patent application. He had 38 American patents (and 53 total patents) that were of the highest quality in that most of them became successful products that advanced an industry.

In avionics, Fred invented the computer for the SHORAN bombing system, the first radar bombing system. Its predecessor, Norden Bombsight, enabled World War II air squadrons to bomb enemy territories and submarines, but it was inaccurate; Fred's invention offered superior accuracy. He also invented the first autopilot for airplanes and the radio direction finder that all American-made commercial airplanes used for 30 years. In the automotive realm, he licensed to

Ford a rear-axle system patent that the company did not end up using but that became vital in General Motors' front-wheel-drive designs. Fred also supervised advanced engineering for the Ford Falcon, Thunderbird, Fairlane, and Galaxie automobiles. And in publishing, Fred invented a digital phototypesetter that almost all newspaper publishers used for many years.

As an engineer, Fred not only invented, but also developed new components, technologies, and equipment. He had no pride of ownership, no not-invented-here syndrome, nor any inclination to say, "I didn't invent this, so I won't engineer it." Instead, he possessed wide-eyed curiosity, undisputed brilliance, and a driving (almost compulsive) desire to move inventions into useful products. For instance, at the request of Leland Clark, Jr., a pioneering research professor at Antioch College, Fred developed the first commercial heart-lung machine; he independently developed the first commercially successful crystal pickup for the phonograph; and, for a music professor at Dartmouth, he developed the first digital music synthesizer.

Considering his prolific and successful engineering and inventiveness, why is his name not better known? Several things explain his relative obscurity.

First, being a passionate inventor and engineer, and not at all a business person, he gave away most of his inventions. Although he and his wife, Martha, fared very well financially, Fred never accumulated massive capital from his brilliance. He did not generate the net worth to donate vast sums of money and ensure buildings, parkways, and bridges bore his name. In fact, he only agreed to co-found one company, and that co-founding ran contrary to his nature. Graduate students he helped out a bit here and there suggested to him that he take equity in their company.

He demurred. What do I want equity for?

But they insisted their brilliant mentor at a minimum receive equity for his incisive advice, and that stock holding in the Yellow Springs Instrument Company ended up supporting Martha throughout her old age.

Second, Fred never wanted an ounce of glory; he only wanted to turn inventions into products that enabled America to win the war or humans to live better or longer. As you will read, this was his driving desire, a compulsion. Although he had boundless energy, his ego was well-bounded. He was humble about his own contributions to the world. He never spoke with family or friends of his many brilliant inventions, preferring instead to ask others about their areas of interest and knowledge. Meanwhile, in his cluttered workshop, he moved on to his next invention . . . time and time again . . . with unusual speed and energy.

Third, Fred had so many personal interests and hobbies, a polymath for sure, that he could not be bothered to expend his energies trying to make a name for himself. He wanted to do, see, and learn about fun things, not spend time figuring out how personally to capitalize on a project. Out of personal interest, Fred built an exact replica of the last steam locomotive. He subsequently rebuilt a camera so he could make the steam locomotive appear to be a full-sized locomotive sitting at a modern-day Vermont train station. Of course, *Model Railroader* featured the wonderful perspective photograph on its cover. He won *Scientific American*'s First International Model Airplane Contest—among nearly 11,000 entries, his laughably simple entry performed best. He became extremely knowledgeable on the topics of botany and birds. He could, and did, enter any area of knowledge, learning, science, and engineering as a neophyte and came out with an invention, a patent, a developed device, a product, a company, a published article, a contest victory, a lifelong friend, reverence of others, a whole lot of fun, and more.

Here introverted and there extroverted, Fred lived a life of love and wonder. He showed great fondness and affection to the love of his life, his wife, Martha, and he had tremendous interest in his children as they matured and their interests coincided with his own curiosity. That said, those who knew him well say he did not process emotions well. He preferred to avoid and ignore the emotional aspects of life in favor of deeply pursuing the engineering aspects of it. Friends and family agree that this habit of emotional avoidance affected him as a father, a grandfather, and a professional. He largely dealt with strong negative emotions by going down into his basement shop, fiddling around with machinery and physical objects and working on inventions until the emotion had passed. He likely did not consciously seek to avoid emotions—he did not consider emotional avoidance to be a life strategy—but those who had close ties to him noticed that he dealt with negativity by leaving a room to . . . tinker.

Of course, avoiding negative emotions by making things might describe many an inventor and engineer. Fred was an engineer's engineer, an inventor's inventor, a polymath's polymath, the archetypal jovial genius, happily possessed by insatiable curiosity, and this book offers more of his story. The book is not written to impress—he would never have wanted people to be impressed by him—but to inform and inspire others who might have interest in engineering, invention, and making things. In Fred, who passed away in 1985, they have a kindred spirit.

It is hoped that this book on Fred Hooven, who was a great inventor—worthy at least of mention in the same sentences as other Midwestern greats of his era such as Thomas Edison, the Wright brothers, and Charles Kettering—inspires readers to invent, to create, to pursue hobbies with gusto, to be curious, to have and express independence of thought and action, and to realize that tinkering, inventing, and engineering are fun, lucrative, and fulfilling.

✿

Young Man

F rederick J. Hooven was born March 5, 1905, in Dayton, Ohio, to
Claude Caldwell Hooven, an Ohioan, and Pennsylvania-native
Anna Elizabeth Johnson. Claude descended directly from General Nathanael Greene, born in Rhode Island in 1742. Greene's family had helped settle the colony after conflict with the Puritans drove the family and others from the Massachusetts Bay Colony. As a grown man, Greene and his wife, Catherine Littlefield, had six children.

Greene is well known to historians because he served as Commander-in-Chief George Washington's first and chief general. He stood side-by-side with General Washington many times during the Revolutionary War, and Washington said that he trusted General Greene completely in battle. Interestingly, Greene was a Quaker, a denomination that practices pacifism. (Quakers have also always believed that women and men are spiritual equals, and women may speak out during worship.) As he became more militant about fighting British rule, he drifted away from Quakerism and was suspended from Quaker meetings. He became known as the Fighting Quaker.

Post-war, in 1786, Greene moved his family south. One day, he traveled to Savannah for business to visit a friend's rice fields and learn to grow rice. Greene stood in the sun for most of the day, and when he returned home, he became very ill with sunstroke, passing away shortly thereafter. His death was deemed a national day of mourning, with American businesses ordered to close.

That was Fred's father's side of the family—its distant 18th century lineage. Fred's mother, Anna Johnson, was very intelligent and entertaining, and in possession of a strong personality. She is, however, said to have been unable to handle negative emotional situations, shielding herself from them as much as possible. Instead, she neutralized her emotions through drinking and smoking—far too much, it is believed, and indeed she passed away in a fire caused by a burning cigarette. It's possible Fred acquired his difficulty processing strong emotions from his mother.

"He had trouble with emotion," Fred's son, Mike, says about his father. "He told me once he couldn't go to movies because he got too emotional."

Although his mother struggled with addiction, Fred seems to have enjoyed his Dayton childhood. Located near a tributary of the Ohio River, the city sits 55 miles northeast of Cincinnati amid mile upon mile of low-lying agricultural land. It was dubbed the Gem City for unknown reasons, but one hypothesis explained by a *Dayton Daily News* journalist is that in 1845, a *Cincinnati Chronicle* reporter made an observation that included a comparison to a gem:

> The most indifferent observer will not fail to notice Dayton. The wide streets, kept in excellent order, the noble blocks of stores, filled with choice, and of course, cheap goods, and more than all, the exceeding beauty and neatness of the dwellings, you at once mark with a 'white stone,' in a small bend of the Great Miami River, with canals on the east and south, it may be fairly said that Dayton is the gem of all our interior towns; it possesses wealth, refinement, enterprise and a beautiful country.

In 1905, when Fred was born, Americans also knew the Gem City to be a city of innovation. In the early 1900s, people knew Ohio to be an industrial powerhouse, and only 15 years before Fred's birth, in 1890, the US patent office had granted Dayton more patents per capita than any other city. The cash register was invented and commercialized in Dayton in the late 1800s by the National Cash Register company (NCR). The year after Fred's birth, Charles Kettering, a leading engineer at NCR, developed the *electronic* cash register. Kettering would go on to be world-renowned for his inventiveness and would become a close associate of Fred's. And, of course, in the early 1900s, working on their Wright Flyer, the Wright brothers were the phenoms of the town. Although they took their first (very short) flight at Kitty Hawk in North Carolina, they conducted much of their designing and experimenting in fields around town, particularly Huffman Field, a cow pasture eight miles northeast of Dayton.

In 1910, when he was five years old, Fred met the Wright brothers. They had been working on planes for about a decade by then. First, they built the Wright Flyer I. In December 1903, in Kitty Hawk, North Carolina, the brothers made four attempts at flight with it. The wind was strong and cold, but their fourth flight that day lasted one minute before the plane pitched into the ground. Four men on the ground tried to hold the plane down, but the wind overturned it, breaking a wing. By 1905, the year of Fred's birth, they developed the Wright Flyer II. And a few years later, their Wright Flyer III became known as the world's first truly fixed-wing aircraft. As a child, Fred loved to watch them pilot it in figure-eight patterns above the cornfields in Dayton.

Fred initially spied the pilot-inventors in secret, but eventually they figured out that he was watching and invited him to spend time observing them work in the shop up the street from his home. He loved to watch the men engineer their airplanes, and by his mid-teens, he considered the Wrights to be friends, especially Orville. They even gave him a few pieces of the wood from the broken wing of the Wright Flyer I. (The Smithsonian now has the restored plane.) When Fred and his friends attempted to build an aircraft, around age 15, Fred sought Orville's advice. Much later, the Wrights would distribute a "VIP Square" section of cloth from the wing of the original plane to Fred as well. This is a special piece but a bit less personal than the pieces given directly to the 15-year-old Fred by Orville himself.

Early on, Fred showed a tendency to become interested in everything. This curiosity and insatiable engagement with the world, especially its mechanical, electrical, and natural aspects, characterized his life from early on. Son Mike offers examples from much later in Fred's life that speak to Fred's longstanding, voracious appetite for knowledge. Fred and Mike became interested in mushrooms. "I'd go around and look for various types of mushrooms," Mike says. "But Dad learned everything about them. He knew all the species. He knew all the scientific names. I remember another time when I was an old man myself, we were in restaurant on Cape Cod, and my dad got up out of the table and walked over to a guy.

'Are you Raymond Loewy, the famous automobile designer?' he asked.

"That was typical of Dad. Everywhere he went, he had curiosity. He was omnivorous with regard to subject matter. He looked into everything."

As a teen, Fred developed enormous interest in the mechanics, electronics, and acoustics of radio and recording technologies. In "Reminiscences of an Old Hi-Fi Hound," a wonderful paper Fred later wrote, he detailed his interest in these areas. He wrote about the first receiver and transmitter he made, explaining that he bought some of the "radio" parts and built others, assembling the whole thing himself. His receiver and transmission contraption involved rigging up a large "aerial" (an antenna) atop the house, across the neighboring vacant lot, and over the roof of the neighbor's house. He wrote:

> I could get NAA [a major radio facility] at night, and one way to make character with the girls was to bring them up and let them put on the earphones to hear NAA, from Arlington. I should add that boys of 14 could drive cars in those days, because there were no such things as driver's licenses, but they couldn't take a girl out in the car at night without somebody's mother coming along.
>
> Generally there wasn't anything much you could hear in the daytime, but at night you could hear the ships on the Great Lakes, which worked on 500 meters. If you were lucky, there would be something in the message that would tell you where the ship was, and when it signed its call letters you could look in the call book (US Gv't Printing Office, 25¢) and find out what ship it was. One of my friends, a little older than I, got a summer job as operator on a Lake ship.
>
> When transmitters were legalized, I got an amateur license and went big-time with a 1-kilowatt spark transmitter. It had a 25000 volt transformer that rested in a tank of oil about four feet long, with an enormous insulator for the high-voltage lead. Fortunately my parents didn't realize that stuff would have killed me if I had got across those leads, and I was careful. To tell the truth I was always a little scared of that rig. The transformer was connected to a condenser, which was a stack of one-foot-square plats of quarter-inch plate glass, with tinfoil between them, the whole being immersed in a big glass battery jar full of oil. This discharged through a rotary spark gap by way of a high-frequency transformer made of wide ribbons of copper strip wound in a spiral. The primary was only one or two turns, while the secondary had maybe eight turns connected to the aerial. When it was working you could stand with a screwdriver in hand and a spark 4 or 5 inches long would jump into it from that lead-in and the aerial glowed in the dark from the corona. It would blow the girls' minds to press that key while the gap roared and flashed, and the spark jumped into your arm.
>
> This rig was too much for the neighborhood. It would dim the lights all over, and the noise was truly deafening when that spark

jumped the rotary gap. The last straw was when our neighbor (whose tree the aerial was connected to) walked under the chandelier in his living room. He was a tall man, and a spark jumped from the chandelier into his head. He said it didn't really hurt, it just felt like a pinprick, but it made quite a snap, and a flash of light, and it scared the daylights out of his wife. He was a nice guy, but even I realized that the transmitter had to go.

Fred was an adventurous learner, always curious, always building, and he excitedly shared his knowledge with others. Some people considered him to be "out there," as this reminiscence illustrates:

> In 1922 I got a job as a radio editor of the local paper, with a 3" column to fill every day. After about a month I wrote, in two successive days, that one day tubes would sell for less than $1 apiece, and that some day the short waves (meaning less than 200 meters) would be worth more than the long waves, because there were so many more channels available. The editor called me in his office and said I was fired, he couldn't have that kind of wild-haired stuff in his paper, readers had called in to complain.

His tales indicated that he loved adventures in engineering—acoustical, aeronautic, electrical, and more. To hone his engineering prowess, in 1924, he moved to Cambridge, Massachusetts, to attend the Massachusetts Institute of Technology. There, he studied aeronautical and mechanical engineering and ended up graduating in three years instead of the more normal four. During at least one of his summers at home, he indulged his passion for radio engineering by working at DayFan Radio doing what else but . . . designing radio receivers.

Although these were his early experiences with radio-related technologies and aeronautical engineering, there would be plenty more.

✿

Growing Family

Fred's closest friend, Wilbur Kennedy, Jr., had a younger sister whom Fred fancied. Martha Galloway Kennedy had graduated from the Emma Willard School, a women's boarding school in Troy, New York. The eponymous Emma Willard had been a mid-19th century activist in the areas of girls and education, and the school Martha attended was a top women's boarding school. As with most women in the early 1900s, Martha did not go on to college after boarding school. Instead, she worked, and at age 21, she married Fred in Dayton. They moved into an apartment on Cary Drive, and he worked as a mechanical engineer.

Martha, too, had an impressive lineage, descending as she did from Robert the Bruce, who was King of Scots, 1306-1329. A famed warrior who freed Scotland from English rule, he is revered as a Scottish national hero. At least four films, one TV series, one opera, and a video game feature Robert the Bruce, aka Braveheart. He received his Braveheart nickname when, upon his death, the Scots extracted his heart, embalmed it, and placed in a reliquary. A knight, Sir James Douglas, wore it on a chain around his neck; another knight wore the key. Good Sir James, as he was known, was leading a band of Scottish crusaders to the Holy Land to give the heart a tour of the land when the knights had to divert the mission to fight against the Moors in Spain. He died during the battle, but his compatriots recovered the reliquary, and it is now preserved at a Scottish abbey. Beyond the heart, the other visible symbol of Martha's high lineage is a castle in Scotland traceable to her ancestors.

At age 23, Martha and Fred had their first child, and in total they had four children in nine years—1930, 1934, 1936, and 1939. John came first. During his years at Oakwood High School, John never developed the kind of rapport with his father that he reportedly would have liked. John is said to have tried hard to earn Fred's approval, but without success; their personalities did not mesh. Fred loved details and getting into the fact-based nuts-and-bolts of matters, whether products or politics, whereas John liked to speculate about the world in a theoretical and broad-minded manner.

Son Mike says about his brother, "John's interests and proclivities probably extended beyond even those of our father, but in a very different way."

John attended Antioch College in nearby Yellow Springs, a liberal arts college that Fred adored. At Antioch, John met Betty Ann Corcoran, and they married. He attended MIT for graduate school on a Fulbright Scholarship, but ended up transferring to the University of Michigan. Being the eldest son of a great engineer proved to be a difficult act to follow. Instead of engineering, he pursued psychology, practicing organizational development at Exxon and Arthur D. Little. He eventually became a fully licensed psychologist.

Four years John's junior, Peter enjoyed a closer relationship with his father than did John. Their personalities and interests overlapped and meshed well. Once, Fred asked Pete if he wanted to build a car, and Pete said he sure did. Although only a frame, the car, which took many months to construct in Fred's shop, was a stunning feat of engineering for father and son. Peter drove the car around Dayton, even though the cross between a buggy and an automobile had no sides or top.

Younger brother Mike recalls, "I was burning up with jealousy."

After high school, when he moved out of the house and attended Antioch College, Peter's interests began to diverge from those of his father. He became a gifted artist, won a Guggenheim Fellowship, and became a professor at the Maryland Institute College of Art, and then moved to Truro, Massachusetts, at the tip of Cape Cod. He lived in a former church built in the 1800s and had his studio there. In the converted-church studio, Peter painted professionally for twenty years. His nephew, Mike (John's son), says he always had guests and many good times in the summer. Mike describes the household as a "continuous party going from May through September" (high season on Cape Cod) and says the gatherings had fascinating people in attendance.

As an adolescent, third son Mike, like his older brother, John, did not enjoy a close relationship with his father. Fred did not exclude Mike from activities, interests, or conversations. Indeed, the *paterfamilias* pathologically brought people into his orbit—everybody wanted to be close Fred, and vice versa—and Mike and his activities certainly interested Fred. But Fred did not go out of his way to include the adolescent Mike. Mike has surmised that perhaps Fred did not at that time in his life know how to engage a child or adolescent; other than his own children, Fred did not connect often, or ever, with little kids.

Plus, Mike says that he slightly feared his father. Although only about 5'9", the man was a giant to his sons. "When Dad got angry, it was rather unpleasant," Mike says. He explains that Fred never yelled or screamed in anger; rather, he indicated with his terseness and his tone that one had done something he did not appreciate. That made Mike uncomfortable, resulting in a tendency to shy away from his father. Mike did, on the other hand, enjoy a warm and close relationship with his mother.

In high school, Mike did not work hard, but when he attended Antioch College, that changed. "I excelled at math!" Mike says. "This came as a surprise to everyone, including me. There was no sign of a skill or proclivity for math whatsoever until college. The only reason I worked at it in college was that the teacher graded homework and counted homework completion on the final grade. I never previously had done homework. But the teacher said to me, 'You're risking failing because you're not doing your homework.'"

When Mike began applying himself, he found that he loved number theory. In his eyes, this is when he began to earn his father's respect and friendship.

Mike earned admission to Princeton University for a graduate program in math, which in those years is said to have admitted 15 of 750 applicants. However, he did not thrive there—upon reflection, he believes perhaps his immaturity prevented him from enjoying the amazing environment of 1960s-era Princeton math and physics legends. (The Princeton Institute of Advanced Study was affiliated with the graduate school of mathematics; Albert Einstein had done his brilliant work there and the brilliant Kurt Gödel, among others, was there.) He left shortly after enrolling. He did, however, continue to develop an ever-closer relationship with his father, one ultimately characterized by warmth and mutual admiration. Mike believes his father began to appreciate the shared interests and experiences they had; likewise, Mike developed a deeper understanding of Fred as a person, involving himself in areas of interest to his father.

Before moving on to the youngest, Martha, it's worth discussing Fred's progressive views on education. For years, he sat on the local Oakwood school board, but he did not necessarily play the part of the dutiful school parent. Daughter Martha says he occasionally kept his children home from school for the day so he could take them on a field trip to see a plane, train, or automobile he admired. He offered no explanation to the school administrators. Martha says, "It never occurred to him to ask permission or let them know where we were."

He once wrote an article for the *Kettering-Oakwood Times* describing how he wanted the Hoovens to be educated.

> For my part, I don't want my kids to grow up to be obedient followers and conformists, I want them to be mature and responsible and kindly and tolerant. I want them to love life and people and living things. I want them above all to have character and integrity. Somewhere I want some part of their cantankerous, hell-raising, rebellious ancestry to show. I can't leave all this to the schools, I know. But I would like the schools to go along with me.

In particular, Antioch College held a special place in Fred's heart. He taught a weekly seminar in engineering problems there for eighteen years, and he long served on its board of trustees. He loved the Yellow Springs, Ohio, institution for its educationally reformist characteristics, which were many. Here's a short list. Its first president was famed intellectual progressive and educational reformer Horace Mann. It governed itself in a co-op style, which meant that students had a stronger-than-usual voice with the administration. It never accepted the submission of SAT or ACT scores. It has always used an integrated curriculum. It was the first college in the country to have a female faculty member who was an equal to her male counterparts. During World War II, it allowed Japanese prisoners of war in American internment camps to enroll and take classes there. And it was the third college in the country to admit Black students on an equal basis with white students, considered a very controversial stance at the time. Indeed, Antioch is the alma mater of Coretta Scott King, an author, activist, civil rights leader, and widow of Martin Luther King, Jr.

Suffice it to say, Fred supported educational institutions that encouraged radical creativity, independent thinking, and progressive mindsets, and Antioch exemplified his educational mindset.

Now, on to the baby of the family, Martha, who also attended the nearby public Oakwood School. As a girl, she considered her father to be very kind and thoughtful, but also living entirely in his own intellectual realm. She could plainly see he knew a lot about many different things—various branches of science, playing and producing music, inventing, and engineering—but to her, he seemed a little unreachable. She says that from her earliest days, she assumed she would never understand anything he was talking about. Dinner table conversations were dynamic, intellectual, and intimidating, especially as the kids became teens. Fred and the boys discussed science, engineering, and politics, while the two Marthas, mother and daughter, listened and only rarely spoke. Young Martha picked up on cues from her mother—smiling, patient, content—that it was a delight to sit in on these conversations. She therefore felt no resentment; rather, she says she felt a sense of awe.

Martha attended Colby College in Maine, where she majored in math. During freshman week, she met Steve Richardson, a fellow math major. As a Massachusetts native who had attended a Vermont boarding school, he was a through-and-through New Englander. They were engaged by senior year and married in summer about a year later.

III

In 1941, the family moved into a home that all six of them felt was a piece of heaven. Fred and Martha had designed it to be a fully concrete construction, a rarity at the time. Today, builders often construct concrete homes because of the durability, comparatively low carbon footprint, energy savings, noise dampening, and advances in

the attractiveness of concrete design. But in the 1940s, a fully concrete residential home (i.e., not a commercial building) was totally
unconventional. The Hooven parents and kids knew that, at a minimum, no one in Dayton had a single-family home with a structure
made entirely of poured-concrete blocks, and grandson Mike believes
the home was one of the only fully concrete residential, single-family
homes in the country.

When the family moved into the 4,200-square-foot home, they
knew it was special. Full of windows, it possessed a light, airy feel
quite different from the dark feel of typical turn-of-the-century
homes, and not at all what one might expect from a concrete building. Fred, in particular, loved to spend time there. As he transitioned
in the late 1930s from working as an employee for companies to
working as a consultant and independent inventor, he had the free-
dom to set his own hours. These were not the hours of a big company
executive. His children recall that he often went into the shop late
in the morning and returned home mid-afternoon. For months at a
time, he might work very hard, and then for the next few months, he
might be home more often than not. The creative schedule matched
his inventive mind, yet never prevented him from making a fine living. On the days when he arrived home mid-afternoon, he typically
went into his favorite room in the house, the living room, where the
grand piano sat.

Fred loved to play piano, and son Mike fondly recalls sitting
underneath the piano while Fred played. "It was a little, special place
to me."

Near the living room, under the main staircase, was a large closet.
There, Fred had a Fairchild Record-Cutting Lathe, an amazing piece
of machinery used in creating recordings. His Fairchild could cut 16-
and 12-inch glass disks. (Early disks were glass, not vinyl. For exam-

ple, Alexander Graham Bell made his first recording, which simply played the words "Mary had a little lamb," on a glass disk.) Fred used the record-cutting lathe to cut high-quality recordings years before anybody in the general public knew of long-playing records or had heard of such a thing as high-fidelity recordings.

And, of course, Fred did not make just a few recordings he made about seventy in all. He recorded his children interacting, Martha and him talking or entertaining family or friends, and conversations with Martha's brother, Wilbur, who was Fred's best friend. He recorded himself playing piano, quite well, as it turned out, and he recorded a close family friend, Joe Raieff, playing piano. Stationed at Dayton's Wright-Patterson Air Force Base during World War II, Raieff played concerts on the military base and in the Dayton area, as well as playing privately at the Hooven house. Post-war, the gifted musician joined the faculty at the famous Juilliard School and eventually chaired its Piano Department.

Altogether as a collection, the recordings Fred made are priceless.

✱

Inventive Engineer

After graduating from MIT with a major in mechanical and aeronautical engineering, Fred returned to Dayton, where he worked as a junior engineer for General Motors. In the late 1920s, GM was an American auto manufacturer on the rise. In 1903, the Ford Motor Company had launched large-scale commercial auto production with its Model T, and by the 1920s, GM had become a ferocious competitor. In fact, GM would soon surpass Ford in market share and retain the top position for many years, while Ford and Chrysler would take second and third places, respectively.

For a GM plant in the Dayton area, Fred worked on a brake shoe component within a drum braking system. A car's brake shoe consists of two crescent-shaped pieces that line the inside of the brake drum. When the driver presses the brake pedal, the two crescent-shaped liners, which are made of a friction material, apply force onto the inside surface of the brake drum, which stops the rotation of the drum, in turn stopping the vehicle. For 25 years, GM used Fred's brake shoe design exclusively in all of its vehicles—a count of millions.

|||

After a couple of years with GM, in 1929, Fred joined Dayton Rubber Manufacturing, where he designed auto suspension systems—the combination of tires, tire air, coil springs, shock absorbers, and rods and linkages that connect a car body to its wheels. A suspension system must balance good road handling (i.e., performance) with ride quality (i.e., comfort), which can be tricky to get right. As US automotive design and engineering advanced and consumers expected superior performance and comfort, the balance of handling and quality became a holy grail for manufacturers.

The rubber components in the suspension system, which is what Fred worked on, include not only tires, but also the rubber diaphragm in the shock absorber and parts of the constant velocity (CV) joint. High-performing rubber components were vital to a quality automobile, and Dayton sat in Ohio, the global center of rubber and plastics manufacturing. Fred was working on highly engineered systems in a rapidly growing industry and company. His innovations on suspension systems would reappear later at Ford and with General Motors. Indeed, his insights would greatly improve the design of American-made cars.

But first, he would transition into the field of avionics.

III

Fred left the Dayton Rubber Manufacturing Company to join the US Army Air Corps as an engineer, a move that marked the beginning of years of extraordinary work in avionics. His work transitions over the next ten to fifteen years appear to be motivated by a combination of love for aeronautical engineering and radio technologies; the Great Depression—which started in 1929 and ended, technically, in 1933; and the build-up to, beginning, and end of World War II. Between 1930, when he left Dayton Rubber to work with the Army, and 1950, he would receive 14 avionics-related patents.

That said, his stint with the Army lasted only a year because in 1931, the country was deeply mired in the Great Depression, and he lost his Army job. He soon found work as an engineer for American Loth Company.

About a decade before Fred's employment there, founder William Loth wrote a seminal paper, "On the Problem of Guiding Aircraft in a Fog or by Night When There Is No Visibility," presented to the French Academy of Sciences. In the paper, Loth described a system for blind landing of aircraft (the term "blind" relates to the need to use instruments because clouds, darkness, or mist make flying by sight impossible) using an artificial electromagnetic field. The French Undersecretary of State asked Loth to build a prototype, which he did. His system passed an alternating current through a loop of cable surrounding a landing field, producing a varying electromagnetic field that an aircraft could detect. Prototype completed, Loth formed a company (Société Industrielle des Procédés Loth) to develop and manufacture his system; the American subsidiary was aptly named the American Loth Company. For this avionics leader, Fred designed, in 1931, a production-ready blind aircraft landing system.

III

In these days of the early 1930s, with the economy very unstable, Fred wanted and needed, financially and intellectually, to invent things on his own account, in addition to working for others. As such, he had his own, private workshop, which he considered to be his true intellectual home. It likely had few peers anywhere for the combination of the breadth of equipment owned by a solo inventor and the charming disorganization of the place. His children describe the place as they recall it.

"Dad's workshop was a fascinating place for a little kid," says Mike. "We sometimes went there and hung out while he worked." When Mike visited, he typically saw that his father might, say, put down a tool and then spend half an hour looking for it—i.e., disorganization and absent-mindedness, but also that his father had an awful lot of fun, behaving like the proverbial kid in the candy store. Mike and his siblings could plainly see that their father adored tinkering, making, refining, inventing, producing, and all the things associated with being an engineer with creative genius and abundant energy.

Youngest child Martha recalls stopping a few times at her father's workshop in Dayton *en route* from their south side home to the swimming pool on the north side. With all the kids in the car, mom Martha stopped at the workshop to pick up Fred. While he finished tinkering and changed into his swim trunks, the kids wandered around the shop.

Although the Hooven children observed in that workshop the work habits of an independent, inventive, creative man, what they did not do was learn the ropes of how Fred did what he did. To their memory, he generally wandered around and did stuff that was hard for children to understand; Fred was too non-linear, creative, and busy producing prototypes and such to teach them.

Fred also had a basement shop at his home in Dayton, which included basic tools, such as a drill press and lathe. Whichever shop he was in, he was extraordinarily creative and productive. Apart from Fred, Bob Kimes was the only person who spent time at the downtown Dayton shop. A tall, thickly built, square-jawed man, Kimes served in the dual roles of independent inventor and right-hand man to Fred. He and his wife were also good friends with Fred and Martha, and Kimes and Fred had fun together, with Fred often playing the practical jokester to Kimes's straight-man. For instance, one story has it that Fred was in one room working on a project when he heard Kimes start swearing, shouting, and screaming. A few seconds later, a massive handheld metal drill came flying through the wall and over Fred's head, landing on the floor with a loud clang.

"So I lay down and pretended I was dead," Fred gleefully told his kids at dinner that night. He had a wicked sense of humor. "When Bob walked in to see what damage he had done with his temper tantrum, he just about died." Fred got a huge kick out of his antics.

|||

In that workshop, on his own account, while employed by American Loth, Fred independently invented and developed the first successful high-fidelity crystal phonograph cartridge. Also called a "pick-up," the crystal phonograph cartridge is an electromagnetic transducer used for playing records on a phonograph turntable. Long-play (LP) records became all the rage in the 1930s and stayed popular well into the 1970s. The pickup's tip, or needle, moves along the serrated grooves in an LP. The movement of the needle vibrates a cantilever containing a magnet that moves between sets of coils. The coils generate an electric signal that's converted into sound. The music or words are then amplified by a speaker.

The first crystal phonograph pickup came out commercially in 1925 and used a bulky, low-fidelity magnet. By the 1930s, the piezo-electric pick-up had been created, but it had not been designed and developed for commercial production and sale. Fred wrote that, to devise his invention, he studied "the classic paper by Maxfield and Harrison, who had done the work on the ill-fated Orthophonic, and began to understand the theory of pickup design." He then sought out some friends who had started the Brush Development Company and who had produced the first piezoelectric crystals that could be used as audio transducers. He used their knowledge and equipment and developed the world's first high-fidelity piezoelectric pick-up that could be used for commercial production and sale.

This invention was widely used for many years. Fred wrote, "Every [jukebox] from 1932 up until about 1952 had one of those pickups in it . . . most of the commercial radio-phonographs used it, too."

Simple, elegant design and the further development of pre-existing inventions was a particular skill of Fred's, and that showed up in the crystal pick-up. His written account of selling the patents relating to it illustrates his lesser skill at negotiating in business and monetizing his brilliance. He wrote:

> In 1940, I had an inquiry from RCA Victor to buy some patents from me. I quoted them what I now realize was a bargain price, but stipulated that I also required free access to the NBC transcription library, which I still fondly imagined contained all these immortal performances of the past decade, recorded with high fidelity and no scratch.
>
> The patent attorney turned a little pale, but manfully went back and started negotiations. I had imagined that RCA and NBC were so closely coupled there would be no need for formality, but I later learned that my request had had to go all the way up to the Board of Directors of NBC. If I had just had sense enough to realize how valuable those patents were, I would have got a decent price for them.
>
> After a long delay, the list of available transcription records began to come in. The list itself would have just about filled a pickup truck. I was flabbergasted, and began to dig through it. In all the sheets I looked at, there wasn't a single disk that had a continuous recording for the entire 15 minutes of play. There was not one single example of a performance of any piece of music worth listening to. It was all just a collection of short "spots." In discouragement, I told them to forget about the records and give me an extra $5000 instead.

III

When American Loth Company folded soon after he joined, Fred got a job as chief engineer at Radio Products Company, where he designed and developed electronic test equipment. A small, Dayton-based company, Radio Products sold the Dayrad brand of tube checkers and measurement devices. With the Depression raging, Radio Products folded, at which point Fred wrote, "I did what odd jobs I could to keep body and soul together, with help from my wife's and my family."

Would that everyone could be so productive with their "odd jobs"—this is when Fred did his seminal work on a breakthrough radio direction-finder for aircraft and much more. His first flagship patent, the radio direction finder, also called a radio compass, and later called an automatic direction finder, was a beacon system that defined an airplane's position lines through emitting signals that could be detected by receivers. The invention advanced aerial navigation tremendously in that it permitted distinction between forward and backward direction, unknown in previous aerial navigation. Prior to the invention of radio direction finders, pilots relied on either celestial navigation or dead reckoning for finding their way from one point to another.

Fred was not the first or only person to invent a radio direction finder; rather, he was the first person to construct a radio direction finder that could be manufactured successfully and remain in service in a plane for years. Because of the combination of advances that Fred's invention and design embodied, many US military aircraft and every commercial aircraft in the world used it for many years, into the 1960s. Fred would much later say (in a 1982 interview): "It was my own idea, and it completely dominated the scene for that kind of device for a time roughly corresponding to the life of the DC-3. It made it routine to cross the ocean, where it had been an adventure before." The DC-3 was famous for the extraordinary upgrades it embodied in air travel, pioneering new air routes and popularizing commercial (vs. military) air travel.

Fred quickly licensed the radio direction finder to Bendix Aviation, and as head of the newly created radio products division, he then acquired the defunct company, Radio Products, to serve as the Bendix base of operations in Dayton. After Fred licensed his invention to Bendix, the Hooven Radio Compass became known as the Bendix Radio Compass, and so the two terms are synonymous.

Based in Indiana, Bendix was technically a good fit for Fred. The company was gearing up for its eventual wartime domination of military avionics. One source states that in World War II, Bendix made *nearly every ancillary instrument or piece of equipment for military aircraft*; another states that during the war, it manufactured about 75 percent of all avionics in US aircraft. It established its radio products division for the sole purpose of making radio transmitters and receivers for aircraft as well as other types of avionics. Bendix also sponsored the then-famous Bendix continental air race, which started in 1931 and was a transcontinental US point-to-point race meant to encourage the development of durable, efficient aircraft for commercial aviation. Fred must have loved being associated with this through his employer.

As background, it's important to note that 1933 (when Fred invented his radio direction finder) is the year Adolf Hitler became Chancellor of Germany; it's the year when US citizens and government alike became alarmed by the rise of Nazism and persecution against Jewish people. American newspapers reported frequently and in frightening terms about the rise of Hitler and Nazism and the devastation to Jewish citizens and culture. Only beginning to emerge from the nadir of the Depression, the US government developed extreme agita about the impact of the rise of Nazi Germany. American powers-that-be knew they needed to build up military capabilities, and World War I had informed the US that winning a war in the air would be of the utmost importance.

With alarm about Nazi Germany being the order of the day, and knowing air superiority was vital, it's reasonable to assume that Fred had an acute sense of urgency and mission about his inventions. Throughout his professional life, he constantly had things in motion . . . many irons in the fire . . . plates spinning . . . and balls in the air. But during the 1930s and '40s, his inventing, developing, and producing activities are especially nonlinear and tricky to follow. A CV he created lists many inventions and developments, and his typed-up accomplishments for the 1930s and 1940s periodically end with either "and many other products," or "etc.," as if there were too many products, inventions, and collaborations to note, but . . . the reader of the CV can get the general idea.

In his CV, Fred does not clearly outline the various roles he played across his inventions—inventor, engineer, producer—and he does not precisely describe the agreements he might or might not have had with others, from working on his own account to employment with companies, and from collaborations with big companies to loose agreements with sole entrepreneur-founders. He does not summarize the various financial arrangements he forged, whether investing his own time and money entirely on spec or having the benefit of a known licensor or investor. This intellectual and financial fluidity and flexibility applied even to Fred's physical act and space of inventing, in that he flexibly operated in various physical spaces. He surely loved his professional workshop, but he also had a workshop in his home, where he often worked. A grid or spreadsheet of his inventions would be antithetical to how Fred thought and operated; and soone is not created here.

Further, with American and Allied military and air superiority being the most urgent effort on the planet, numerous of Fred's inventions and developments are assumed to be classified. This means they were explicitly not submitted into the patent canon. The role of a patent is to inform and teach, which was wholly contrary to US technology directives in the late 1930s and throughout the 1940s. The US did not want enemies to have access to American technologies. No one could search for patents online as we do today (there was no internet), but with the patent database legally available to the public, enemies could access US patents.

With that being said, at Bendix, Fred collaborated in the development of the automatic steering system—a.k.a., autopilot—for aircraft that was used in "the first unmanned flight by the US Army Air Corps in 1937." This is presumably the same flight that is excitedly described as keeping "a US Army Air Corps aircraft on a true heading and altitude for three hours." This initial system went on to have improved control algorithms, hydraulic servomechanisms, and radio-navigation aids, all of which made it possible to fly at night and in bad weather. Although Fred had left Bendix by the time it made those later developments, he was involved in components used in the later autopilot upgrades.

✲

Independent Inventor

After two years at Bendix, Fred left. One source says he had a falling out with founder Vince Bendix about the founder's "overzealous business practices." This is when Fred kicked off in earnest his long stint of exclusively independent inventing. No longer would he, as he had heretofore done, join companies as an employee only for the company to go bust or for him to depart in frustration. He clearly loved independent inventing, and starting in 1936, he would be for the next 18 years, an exclusively independent inventor.

It was unusual in the 1930s, '40s, and '50s to be unaffiliated with a corporation. Most people believed only the most unemployable people took this path. But Fred marched to the beat of his own drum. He thought and acted independently, and as important, he had fierce loyalty not to any single company—that was never his unit of interest—but instead to science, progress, and engineering. He never sustained loyalty to a corporate modus operandi or higher-up over a loyalty to science, progress, and engineering. And so, his acting as independent-minded, unaffiliated consultant and inventor allowed him to maintain that loyalty.

In his professional and home workshops, he became ever more enmeshed in inventing products related to radio waves. The suite of radio-related patents he created became integral components of systems for radio-based navigation. They are:

- Direction Finder (f 1934, g 1936)
- Radio-Controlled Aircraft System (f 1936, g 1939)
- Control Mechanism for Radio Receivers (f 1936, g 1939)

- Radio Compass System (f 1936, g 1940)
- Radio Compass (f 1936, g 1940)
- Radio Receiver (f 1936, g 1939)

- Radio Compass (f 1940, g 1942)
- Radio Compass (f 1936, g 1943)

*f = filed, g = granted

Additionally, Fred invented aircraft landing systems. The patents include an air navigation and landing system (submitted 1936, granted 1939) and an aircraft landing system (submitted 1936, granted 1939). Like the radio compass, these are categorized in the patent canon as "beacons or beacon systems transmitting signals having a characteristic or characteristics capable of being detected by non-directional receivers and defining directions, positions, or position lines fixed relatively to the beacon transmitters; receivers cooperating therewith using radio waves." Of his blind aircraft landing systems, Fred later wrote:

> The first blind landing was made by Jimmy Doolittle in 1929 and he did not use a direction finder. The first blind landing system that did not require a genius to operate it was made in 1934 using an experimental device invented by Geoffrey Kreusi. My direction finder was used on the first fully automatic landing which was accomplished in 1937 at Wright Field with a manned aircraft and later that same year at Muroc Lake in California with an unmanned aircraft.

III

And, Fred invented bombsights and bombing mechanisms (the bomb-holding structure and the electronic bombing intervalometer) that comprised a bomb-dropping system. The intervalometer was a device that enabled an aircraft to precisely time (hence the words "interval" and "meter") the sequential release of its bombs, thus avoiding catastrophic failures. A military plane in the 1930s and '40s typically released 20 bombs per second. A catastrophe could be defined as a bomb failing to release, which sometimes occurred, with the next one releasing on top of it, exploding the plane. The intervalometer Fred invented was electronically and mechanically advanced such that if a bomb failed to release, the bombing system would stop. His patents in these areas included a Timing Device (1939 and 1942), Bomb Release System (1941), and Bomb Dropper (1941).

These inventions constituted the computing system for the groundbreaking automatic SHORAN bombing system, the first radar bombing system. His contribution to the SHORAN included the intervalometer, the bomb rack release and control system, and the relay logic for bomb rack automatic selector. Some portion of these patents also ended up being used as the first electronic gun-control servo, which senses errors and then corrects the action of the gun.

The SHORAN bombsight replaced the predecessor Norden bombsight. As the bombing system the Allied forces used for most of World War II, the Norden consisted of more than 2,000 parts such as DC motors, gears, potentiometers, clutches, levers, cams, mirrors, gyroscopes, and specialized components. In order to boost troop morale and put enemies on notice about technological superiority, the US touted the Norden bombsight as an incredible bombing system. The technology was highly classified, its details not in the form of a patent, which by definition would open up the technology to the public realm and risk it getting into enemy hands. Instead, its specifics were in the form of a closely guarded government secret. Since it was one of the country's most important military secrets, no photographs or release of specifications or performance data to the public were allowed.

In reality, however, it performed quite poorly. One publication states:

> The actual performance of the Norden in combat was good some of the time, but rarely great, and often terrible. Several studies revealed that as few as 5 percent of Eighth Air Force bombs fell within 1,000 feet of the target and the average error for 500-pound bombs dropped in Europe was a whopping 1,673 feet. There are examples of many hundreds of bombs aimed at a single small target with only one or two bombs reaching their mark. Some gross errors were even measured in miles.

Although absolutely vital in helping bombardiers achieve the Allied objective of destroying enemy infrastructure, the fact is, the Norden bombsight system suffered from low precision. The government prioritized producing the Norden bombsight so that Allies would have *something* to use for bombing missions, and the military-industrial complex urgently produced it. At the same time, however, the government put an elite team to work on a more accurate replacement—the SHOrt RAnge Navigation, or SHORAN system. SHORAN used ground-based transponders to respond to interrogation signals sent from the bomber aircraft. By measuring the round-trip time to and from one of the transponders, the distance to that ground station could be accurately determined. The aircraft flew an arcing path that kept it at a predetermined distance from one of the stations. The distance to a second station was also being measured, and when it reached a predetermined distance from that station as well, the bombs were dropped. SHORAN could guide dozens of planes, limited only by how rapidly the ground station's transponders could respond. Basically, SHORAN served as an electronic navigation and bombing system that used a precision radar beacon.

Years after the SHORAN helped Allies win the war, the *Dayton Daily News* wrote an article about Fred's role in inventing it. The occasion was the Air Force Museum commissioning the invention into its collection. In the article, Fred talked about his involvement in the initial government tests to ensure it had greater accuracy than the Norden Bombsight.

> Hooven said first tests were disappointing. Bombs, dropped on a tiny rock island off the Florida coast, landed 600 feet from the target. On the second trial, bombing from a different direction, they hit the same spot. The same with the third.
>
> Finally it was discovered the computer was putting the bombs exactly where the charts said the rock was located. An unsuspected error, due to variations in local gravity, had made the same mistake in all of the charts of the coastline.

It turned out that Fred's SHORAN computer did not make inaccurate calculations. Rather, cartographers had made their maps inaccurately. In any case, the SHORAN turned out to be much more accurate than the Norden bombsight. The government kept the SHORAN system a highly confidential secret and would only later allow inventors to add it to the patent literature. The Allies outfitted the Martin B26 Marauders with the SHORAN system and set up stations on the Mediterranean front (Corsica, Italy and Dijon, France) in the latter part of the war. In "The History of Shoran," a chapter in *Photogrammetric Engineering & Remote Sensing* , H.G. Sennert writes:

> In September 1944, the first Shoran systems were shipped to Corsica for use in bombing vital pin-point targets in Northern Italy. Following a brief period of organization and training, the first tactical Shoran mission was flown 11 December 1944. From that time on, success of the project was assured. In rapid succession, targets that had been near misses by visual bombing methods were destroyed by means of Shoran. Further losses of airplanes and

crews due to enemy action were sharply reduced as missions could now be scheduled at night or in overcast conditions.

SHORAN then saw consistent use in the Korean War in B-29s. In order to avoid enemy MIG-15s, American planes needed to stealth bomb at night, not during daylight hours, and only the SHORAN system worked well in darkness. The Air Force feverishly added the SHORAN bombsight system to all B-29s. By 1952, the military had installed SHORAN in almost all American fighter planes. Its superior accuracy is cited as one reason the Koreans essentially ceded the war and Korea signed an armistice with the US. In the 1960s, the US military switched to long-range bombing using nuclear weapons, thereby obsoleting SHORAN for military bombing purposes.

Fred's résumé chronicles some of his SHORAN-related inventions as follows:

> . . . Inventor. Development of bombing intervalometer; bomb-rack release & control system; relay logic for bomb-rack automatic selector system; gun-control servo; magnetic recorder precision timer; computer for SHORAN bombing system (the first radar bombing system); automatic SHORAN bombing system.

III

During the first year of Fred's 18-year run of independent inventing, he had a fateful interaction with someone he admired—Amelia Earhart. Earhart had long been a different type of woman. As a girl, she had played basketball and had taken an auto repair course. Watching pilots take off, land, and conduct aerial maneuvers during World War I, Earhart developed an interest in flying. She took her first flight in 1920 in California with a former military pilot. She loved it and began flying lessons. Within months, she had bought her own plane. By 1932, she was racking up "firsts" in aviation—the first woman to fly solo across the Atlantic Ocean; first solo, nonstop flight across the US by a woman; and first person to fly solo from Hawaii to the mainland US. Soon enough, she set her sights on becoming the first person to circumnavigate the globe.

She and navigator Fred Noonan needed to outfit her twin-engine Lockheed 10E Electra with the best instruments in the industry. In 1936, Earhart landed at Dayton's Wright Field and met with another American passionate about aviation, Fred Hooven. It's unclear whether Fred met with Noonan as well—one account says they did, but Fred's own words indicate otherwise. He wrote, in a later report: "I met Miss Earhart for lunch at Wright Field in the summer of 1936. She was accompanied by a younger woman flyer, quite unknown at the time, Jacqueline Cochran." He made no mention of Noonan.

Whatever the case, Earhart and Noonan were eager to have the amazing new radio compass (invented by Fred and produced by Bendix) installed in her plane. Fred met with Earhart and ensured the radio direction finder was installed (from The International Group for Historic Aircraft Recovery): 'A Bureau of Air Commerce inspection dated November 27, 1936, describes a Bendix Radio Compass 'rear of copilot's seat in cabin' (as shown in the photo)." Earhart and her cohort stayed one or two days and, with the Bendix equipment installed, they departed a day or two later. Fred said that he enjoyed his interactions with Earhart and considered her a friend.

In June 1937, Earhart and navigator Noonan took off from Oakland, California, on their famous eastbound flight around the world. They flew to South America, across the Atlantic to Africa, then east to New Guinea by June 29. From there, they took off and aimed for Howland Island. However, during that flight, the plane lost radio contact with the US Coast Guard, and after a two-week search, the Coast Guard declared Earhart and Noonan lost at sea.

The US government took the official position that Earhart and Noonan had crashed into the Pacific Ocean. Most Earhart experts believe that they became lost and ran out of fuel. Fred must have agonized over the possibility that the radio compass in which he had so much confidence had malfunctioned and caused them to lose their bearings. But his presumed agony over his own possible role in her accident was relieved when he found out that the radio compass he had helped install in Earhart's plane was removed very shortly before she left. Noonan and Earhart had made the decision to replace Fred's radio compass with a lesser system, also made by Bendix.

Here are excerpts about the radio compass from The International Group for Historic Aircraft Recovery (TIGHAR), dedicated to promoting responsible aviation archaeology and historic preservation. In the organization's Earhart Project, group members write extensively about the Earhart disappearance mystery:

> Hooven later alleged that Earhart ditched his system for a more primitive system in order to save weight. His claim is corroborated by photos that show that the distinctive faired-over loop of the Hooven system disappears in early 1937 about the same time the loop first appears. The Bureau of Air Commerce inspection done when the airplane came out of repair on May 19, 1937 makes no mention of a second receiver.

and

> Hooven's DF (direction finder), which operated on the conventional low frequency bands, featured a small loop in a low-drag streamlined housing, and though the original design circuits were deemed unreliable by operators at the time, the system would eventually be made automatic in its operation, and as the "ADF" (automatic direction finder), would become the de-facto standard for commercial aviation for many years. Unfortunately, perhaps, the Hooven system was removed from the Electra soon after installation and replaced by another prototype which Bendix people were

hoping to sell to the U.S. Navy. It purported to utilize high frequency (3-10 megacycle) radio waves, especially 7.5 megacycles, corresponding to the amateur's cherished 40-meter band. Apparently Earhart and her husband, promoter George Putnam, were led to believe it was a magic device. It wasn't, and Lawrence Hyland, who was a Bendix vice president at the time, later denied it was aboard when the Electra disappeared.

I'd reason that most likely the swap was made was to preclude Hooven's state-of-the-art ADF from falling into the hands of the Japanese in the event of the Electra coming down in their territory.

and

It looks, to me, increasingly like Earhart removed the Hooven Radio Compass, which entailed a separate receiver and a sense antenna, and replaced it with the new Bendix loop and coupler which used the existing WE20B receiver and did not employ a sense antenna. The belly antenna was the receiving antenna. When it was lost on takeoff at Lae, Earhart lost her ability to receive until the one brief moment when she "switched over" to the loop and heard the "A"s on 7500. She then switched back to the missing belly antenna and again heard nothing.

Because of his friendship with Earhart and profound interest in the fate of her flight, Fred much later became involved with communities that specialized in researching the Earhart disappearance mystery. He corresponded at length with Frederick Goerner, a San Francisco-based radio broadcaster and leading Earhart researcher. In 1966, Goerner published the *New York Times* best seller, *The Search for Amelia Earhart*. In the 1980s, the two Freds became close friends, and Fred Hooven devoted much time to combing through Goerner's writings and talks, seeking the solution to the Earhart mystery.

Fred Hooven decided he believed the post-crash radio messages intercepted by Pan Am direction-finding stations in the Central Pacific were authentic and thus that Earhart had never crashed but somehow landed on an island and used her radio to call for help.

Fred told Goerner he believed one of the intercepted calls from the plane offered evidence of a radio that had a recharging battery. Fred surmised, based on his calculations, that Earhart ended up on McKean or Gardner Island (also called Nikumaroro). He believed the Japanese recovered Earhart and Noonan and took them and their plane to Saipan—this explained why no one ever found any trace of her plane or crew on either island.

To share his theory, Fred authored a 1982 report, "Amelia Earhart's Last Flight," that was both respected and controversial among Earhart specialists. What they greatly respected was Fred's precise calculations, his creativity, and his scholarly approach. He delivered it at a conference at the Smithsonian National Air and Space Museum Library in Washington, DC, and it remains archived there. In the report, now commonly dubbed "The Hooven Report," Fred concluded that the evidence strongly supported the hypothesis that the flyers landed on one of the two islands and transmitted signals during the next three days. In the report, Fred did not tackle the issue of the "plane in the water" (a splash was at some point detected), but instead assumed a wheels-down landing on firm ground so that one of the plane's motors could be operated briefly for battery charging. Of the report, Earhart scholar Cameron Warren concluded that, "Despite a minor flaw or two, Hooven's contribution to Earhart research is substantial, and given his scientific background, extremely valuable. Had he lived longer, he truly might have 'found' Earhart.'"

What was controversial was the entire hypothesis Fred espoused. One Earhart enthusiast commented on Fred's hypothesis that ". . . the idea of the Japanese mounting a short notice operation to sail past/ through the massive U.S. Navy search to essentially kidnap [Earhart] and [Noonan] as well as remove the Electra and all traces thereof from Gardner Island, and get away undetected, stretches all credulity. How an intelligent human being such as Fred Hooven could have come up with that idea is simply beyond me."

Fred later agreed with this critique. His subsequent correspondence with Goerner convinced him that he (Fred Hooven) had been wrong in his own Hooven Report. Goerner showed Hooven that many people lived on the two islands where Fred surmised the Electra had landed, such that the possibility of Earhart landing undetected was exceedingly low. Further, the likelihood of the Japanese getting a ship into the area undetected by the US Navy, which was conducting a massive search, was very low.

Fred never had a problem conceding to great arguments made with facts and calculations. He always thought trying and failing at something was far better than never trying at all.

|||

Fred worked on the (automatic) radio direction finder and landing and bombing systems into the early part of the 1940s. Then, in the mid-1940s, nine years into his eighteen-year independent inventing era, he began to help four Dayton-area men submit a bid for an RFQ that had come from Dayton's Wright-Patterson Air Force Base. The young men needed his insights into the mechanics and electronics of an instrument. They found Fred's contributions to the creation of the product they had outlined in their bid to be so helpful that they insisted he become the fifth stockholder of the company they founded, Yellow Springs Instrument Company.

"He was just helping some students," says Fred's grandchild Mike Hooven. (Fred's grandchild Mike is the eldest son of Fred's eldest son, John. Henceforth, this book will refer to the elder Mike as "son Mike" and to the younger Mike as "grandson Mike") "That's how he thought about it. He helped them with some ideas they had. They were primarily his ideas, but that wasn't important to him.

They said, 'Well, we've got to pay you,' and he said, 'No, don't worry about it.' They paid him with stock that at the time wasn't worth anything."

The year of Yellow Spring Instrument's founding was 1948, and Fred barely took note of the stockholder certificates. He did not exactly crumple them up and throw them under the bed, but he certainly did not pay them any heed. The certificates sat around and turned yellow with age, available somewhere in his messy files. After a few years, the company developed a usable instrument that measured the stability of liquid suspensions, and then it developed temperature probes and instruments. It shortened its name to YSI and became known as a top company providing water monitoring and sampling solutions. As of this writing, it develops and sells products that provide data for better understanding and management of water resources (wastewater process control, climate change and drought studies, flood monitoring and warning, stormwater runoff monitoring, groundwater quantification and contamination, etc.).

As for the long-forgotten, little-noted stock certificates . . . after Fred died in 1985, they held enough value that Martha sold them. The proceeds of the sale supported her for a decade.

III

In the late 1940s, Clarence Lapedes contacted Fred for help on an invention. Lapedes's father, William, had started selling clothing to farmers in the Dayton area in 1896. By the time Clarence contacted Fred, the clothing company, called Lion, had made a name for itself as a leading designer and maker of durable, harsh-workplace clothing that featured advances in clothing design and construction. Today, Lion Apparel sells items related to harsh work environments, has many patents, and states that it is the only clothing company with a dedicated research and development unit.

In the late 1940s, Fred invented for Clarence an electric temperature control system for an electric blanket. Fred's temperature control system maintained consistent, accurate heat by sensing the actual temperature of the room and delivering feedback to the blanket's heating mechanism. He designed it to be so that there could be no adverse effect when a consumer accidentally broke the heating unit. Further, he designed the electric blanket heating system to be simple to construct and inexpensive to produce. We do not know whether or how this patent that Fred filed and assigned to Lapedes was used by Lion.

III

In the 1950s, during the golden age of print newspapers and magazines, Fred helped advance the printing industry a great deal. He did this by inventing a digital electronic phototypesetter that most newspapers ended up using for decades. For 500 years, printing technology was fundamentally based on the Gutenberg press, which had two main aspects: the typesetter, with which type was set in metal, and the printer, also called a lithograph, which used a roller to press the inked metal type onto paper. In the 1900s, the offset printing revolution took place. It replaced the mechanical typesetting process using actual metal type with a mainly chemical process that transferred ink to paper. Phototypesetting, as opposed to mechanical typesetting, allowed typesetters to use film images of characters (letters, numbers, etc.) to expose photographic paper directly, resulting in pages of typeset text. Phototypesetting involved no metal characters at all: the clear and powerful benefit was the ability to print fonts and graphics at any desired size, as well as faster page layout setting.

Very quickly, printers and publishers wanted phototypesetting to become digital and electronic, and the Harris Automatic Press Company intended to be a first mover in the space. Headquartered in

Cleveland, Ohio, Harris was a leader and innovator in the production and sale of commercial printers. To innovate in the digital typesetter space, it made two main moves. First, it acquired the Intertype Corporation, a Brooklyn, New York-based typesetter, and renamed itself Harris-Intertype. Second, it brought on board Fred Hooven, who would help develop the technology that would enable Harris-Intertype to lead in the hot, new space. For Harris-Intertype, Fred submitted (as co-inventor) this patent:

> Photographic Type Composition
> An [object] of this invention is to provide in photographic type-composing apparatus a computor which receives spacing information corresponding to the spacing requirements of the individual successively projected characters and actuates a transducer which mechanically shifts a deflecting means interposed between the character projecting means and the film record strip upon which the images are recorded to shift the axis of each image projected on the momentarily stationary record strip through a desired number of small discrete space units, and which computor totals the space unit information supplied to the transducer and actuates a suitable record strip advancing mechanism to advance the strip through a large space increment equal to an integral number of space units whenever the total number of space units is at least equal to a single large space increment. This invention relates to photographic type composition, and particularly to an improved system for spacing selected characters on a photographic record film as such characters are initially composed into record copy.

The digital electronic phototypesetter was known as the Harris Fotosetter. An entry in Wikipedia describes the Fotosetter as "one of the most popular 'first-generation,' mass-market phototypesetting machines. The system is heavily based on hot metal typesetting technology, with the metal casting machinery replaced with photographic film, a light system and glass pictures of characters. This machine was the Harris-Intertype top seller for many years."

Related to the glass pictures and the characters, Fred once told his son John that he conceived of the digital phototypesetter "while dreaming of people dancing on the water." Presumably, the type characters and the chemical fluids of phototypesetting somehow made Fred think of human characters flitting about on water.

III

Through his involvement with Antioch College, Fred became friends with a biomedical engineer, Leland Clark, Jr., and this friendship accounts for the fact that one of the last items Fred developed during his independent inventing was in the biomedical field. In addition to counting three of his children as Antioch graduates, teaching there, and serving as a trustee, Fred also served as a volunteer researcher at its Fels Research Institute for the Study of Human Development. (He also would eventually count a daughter-in-law and a grandchild as graduates.) For his part, Leland Clark taught biochemistry at Antioch, oversaw research at the Fels Institute, and invented prolifically. They knew each other well.

When it came to science, the two men were kids in a candy store. Over his lifetime, Clark earned 80 biomedical-type patents, and he once told a reporter that when he started high school and discovered that science was an academic discipline, complete with course work, lab sessions, and grades, "It was like discovering that you could get a grade for eating chocolate ice cream."

Clark knew Fred to be brilliant not only at conceiving of things and inventing them, but also at building things. In the 1950s, cardiac science made huge advances because surgeons returning from World War II battlefields intended to make progress on what they had observed in the field. Although physicians could diagnose heart problems better than ever before, surgeons needed better ways to fix the problems. In this light, Clark was working feverishly on oxygen dispersion and sensing in blood, a key aspect to enabling open-heart surgery. In 1950, he published a paper about his inventions, but he needed to test it in operating rooms during open-heart surgery. Only a very few open-heart surgeries had been performed, with most patients not making it through. Why? Because the crude heart-lung machines did not work well or reliably enough. A main problem was that the machines lacked the ability to sense oxygen levels in blood and consistently and rapidly infuse oxygen into the blood.

Clark asked his friend Fred to help him develop a heart-lung machine that would employ Clark's patented oxygen diffusion and sensing technologies. The oxygen sensor represented the breakthrough component of the heart-lung machine—it is considered the first-ever biosensor, and many consider it one of the most revolutionary devices in medicine of the past 50 years. For the electrode invention alone, Clark is considered the "father of biosensors."

The machine Fred developed for Clark used the breakthrough technology and, to boot, could be completely disassembled and sterilized. In the 1951 paper announcing the development of the successful machine, "A Large Capacity, All-Glass Dispersion Oxygenator and Pump," Hooven, Clark, and third co-author (Frank Gollan) wrote:

> Blood is oxygenated by tiny bubbles of oxygen dispersed in the blood. These bubbles are coalesced on a polymethylsiloaxane surface and eliminated in a bubble trap which also serves as the pumping chamber . . .
>
> Previous work has demonstrated the high efficiency of dispersion oxygenation as a means of substituting for the function of the lungs. The present report is confined to the description of an improved, simplified, all-Pyrex apparatus. Pumping is accomplished by an alternative suction and pressure controlled by two electrodes, suspended in the chamber, which actuate two solenoid valves. The apparatus causes no significant damage to the red blood cells. The apparatus is portable, easily cleaned and sterilized, and inexpensive

. . . The blood volume of the apparatus is constant except for the stroke volume and in this respect resembles the normal heart.

For one of the country's most famed medical inventors, Fred delivered! He figured out how to build a machine that combined an oxygen delivery method, a sensor, and a pump. The heart-lung machine was superior to the devices used before it, and future developers of heart-lung machines would rely on its fundamental technology and design. The oxygen sensor remains a core component in every heart-lung machine in every hospital operating room in the country. Today, doctors perform 750,000 open-heart surgeries every year using heart-lung machines that have oxygen electrodes as vital components.

The two men's respect and care for each other remained strong throughout their lives. For instance, as a board member of YSI, Fred ensured that the company licensed and sold inventions made by Clark. Still, the most powerful Clark-Hooven collaboration involved the invention by Clark of the componentry for, and then the development by Hooven of, the complete heart-lung machine. Their collaboration vastly improved the practice of medicine for the benefit of human health.

✿

Family Man

Between Martha and Fred, it is said that Martha had the lion's share of emotional intelligence. She had a soothing, warm presence and a relaxed style. Her intuitions as wife, mother, and homemaker were wonderful, and those who knew her well say she never said anything negative about her husband or even appeared to think anything negative about him. She seemed at all times to enjoy that her husband was eccentric yet very capable. The attitude her warmth and smile embodied about her husband's eccentricities was, "Oh, that's Fred!"

Fred considered Martha to be the apple of his eye. Throughout his marriage, he appeared to others to be infatuated with her, in love, and romantic. His bigger-than-life personality became smaller, more contained, and more deferential in her presence. "They were really a team," Dartmouth engineering colleague John Collier says about the couple. "They treated each other like best friends and lovers, a perfect match."

As to whether Martha fully appreciated the depth and breadth of Fred's intelligence, no one knows. Everyone who knew Fred found him to be brilliant, and surely Martha was no exception, but some people believed he rose to the level of true genius in the realm of Alexander Graham Bell, Thomas Edison, the Wright brothers, and others. Fred never seemed to consider himself brilliant or a genius—he had easy confidence in his knowledge and skills, but he seemed hardly to give passing thought to the unusual nature of his broad and deep capabilities. Rather, Fred enjoyed being with people, making things for others to enjoy, collaborating, and encouraging the growth of people around him; he hardly played the role of introverted savant or awkward genius. Others observed that Martha seemed to know full well that her husband possessed extreme intelligence, but as to whether she considered him a true genius—no verdict. "If she did, she never let on," says Mike of his mother.

What is known is that Fred's inventiveness, playfulness, extroversion, and high energy played out amply in his family life. Here is a handful of simple, nothing-special glimpses into day-to-day family life with Fred.

Daughter Martha says that her strongest memory of her father is his love of trains. "We had a casual-dining breakfast room and a formal dining room in our house. Some nights, we ate in the breakfast room. I can picture sitting at the breakfast room table and Dad was constantly checking the time.

"'Okay, time to go meet the train,' he'd say, and he'd have us all pile into our Buick and head over to the railroad crossing. This was in the late 1940s and early '50s, and the steam engines were really loud. They frightened me. But it never mattered that we were in the middle of eating dinner or had homework or anything because Dad was a nut about trains. He built model trains, he drew trains, he had trains on the brain. Anytime he could see a train, he wanted to see it."

Grandson Mike additionally recalls that every visit to his grandparents' house in Michigan consisted of multiple excursions to a local train crossing to see the steam engines. He says that a good portion of the family was present to see the last commercial steam engine run.

When Martha was older and married to her Colby sweetheart, Steve Richardson, her father helped with the formation of what became a very successful puzzle company. Having been laid off from a technology job in the 1970s, Steve needed to find work that could support his family. With a partner, Dave, he formed a games company called Stave (a combination of Steve and Dave). They decided to enter the puzzles market and came across a tiny, elite jigsaw puzzle company. Steve asked his father-in-law whether he liked the product, and Fred said he loved it!

"I've got a jigsaw in my workshop that you can have for making the puzzles," Fred said excitedly.

Steve says his father-in-law's enthusiasm helped him gain the confidence he needed to enter the puzzles market. "He really helped us decide we wanted to make a go of it."

Son Mike tells a story about how, when he was 15, he asked his father to help him build an amplifier. The task involved first constructing a chassis on which components such as transformer and condensers could be mounted. Fred used a band saw to cut a large square of sheet steel, then a specialized drill to make holes where tubes could fit. Mike admired his father as he watched him employ machines to bend the steel, mount the components onto the chassis, and solder the holes shut. Fred also showed Mike how to develop a circuit diagram and test circuits using an oscilloscope. This was a mini electrical engineering course that Mike found particularly instructive.

Fred loved birds and often went birdwatching. He did not observe birds for peace and relaxation so much as to learn enough about the birds to be able to identify them by call or plumage. He studied birdwatching books and used binoculars, pen, and paper for noting what he saw. Mike went birdwatching numerous times with his father. As a pre-teen, he once joined his father on a trip to Englewood Dam, a local hotspot for birding. At one point, Fred overheard a woman in conversation with someone. He recognized the way she spoke, something about the terms she used and the references she made.

"That must be Edith Blincoe," Fred said to his son. Blincoe authored a weekly column about birds in the *Dayton Journal Herald* that Fred read religiously.

Fred walked over to find her, with son Mike tagging along behind him.

"Are you Edith Blincoe?" Fred asked.

"Yes," she said, and Blincoe and Fred struck up a conversation about birding.

"That was very typical of my dad," Mike says.

Fred found the acquisition of knowledge and development of insight to be fun, so that activities that many people might consider to be normal fun, Fred did not; activities that did not involve knowledge acquisition were, de facto, not much fun. He did, on rare occasions, do things that his children wanted to do simply because they were, to the kids, fun, relaxing, enjoyable" activities. For instance, a few times, Fred took his kids to watch the Cincinnati Reds play ball at Crosley Field.

"Dad was a baseball fan. He often had the game on the radio in our living room. Going to the big city—Cincinnati—was a big deal," daughter, Martha, says, "making the drive there as a whole family and going to a baseball game and the zoo."

Son Mike remembers that once when Dairy Queen came to Dayton, he asked his father to take him for an ice cream. Dairy Queen started in 1940 in the Midwest, and soft serve ice cream in cones became all the rage. As they licked their cones, Mike delighted with the treat, but his father complained, "This stuff tastes like library paste."

Daughter Martha says that another activity Fred was nuts about was car racing. He frequently listened to the Indy 500, and although he never witnessed the race in person, he did find going fast in cars to be a lot of fun. He regularly tested out how a ride felt when he was pushing the vehicle past 100 miles per hour. Grandson Mike recalls a story about the time when Fred owned a Mustang prototype in Detroit. Fred was pushing 100 mph when he passed a police car on the side of the highway. A few miles later, Fred pulled over to the side of the road, and that's when the police car pulled over and gave him a ticket.

"Why'd you stop?" the officer asked Fred. "I never would have caught you."

"I saw a rare bird over there"—Fred pointed to a tree—"and I wanted to get a closer look."

The cool car, the love of engineering performance, the desire to go fast, the focus on nature and birds, the *joie de vivre*, the sanguine approach to getting ticketed—they all fit precisely with what Fred's family and friends know of the man. It's said that Fred had to go to court on an almost weekly basis in order to maintain his license; he loved to drive cars very fast.

❖

Iconic Advisor

In the mid-1950s, Ford executives decided the company needed to reduce the parts comprising Ford automobiles significantly. The unibody initiative became a huge, advanced engineering effort, but when rapidly rising Don Frey asked the company's massive engineering department to make it happen, they told him it could not be done. (By 1964, Frey would be chief engineer, and he would become famous for developing the Mustang.) Engineers explained that the so-called unibody car made no sense. Exasperated, Frey and other executives decided they needed to reach outside the Ford ranks and bring in someone who could tell Ford's engineers how to be more creative.

Because of his love for, and many contributions to, automotive engineering over the years, Fred's name came to the top of the list. It seems the great automobile designer Gordon Buehrig originally put the bug in Ford's ear about hiring Fred. Buehrig had designed the iconic Cord 810/812 automobile, considered one of the most beautiful cars ever designed. It was the first American-designed-and-built front wheel drive car with independent front suspension. It also had pontoon fenders with flip-up headlights operated by a dashboard hand crank. By the time Buehrig was a leading designer at Ford, he had already made a name for himself. Here's what he said about Fred in a magazine interview:

Fred Hooven was my genius friend. One of my genius friends; I've had about three. In 1928 I was working for Packard . . . and Fred Hooven who had just graduated from the Massachusetts Institute of Technology, was on his first job at General Motors research . . . he was just so much smarter than I that I really loved to talk to him. Then later on our paths crossed when I was at Ford and I was given the job of body engineer on the Mark II Continental. I had been down to Dayton to visit Fred and Martha, and Fred had a chassis with a completely new kind of suspension, and he took me for a ride in it. It was just a chassis with no body, and he had built it with his kids. I was so impressed with it that I told my boss at Ford . . . the chief engineer, about it, and arranged for [the chief engineer] and his wife and my wife and me to go to Dayton, and [the chief engineer] took a ride in it and was so impressed with it that he went back to Ford and persuaded them to buy it for its patent. So Fred came in at a high level in Ford Motor . . .

The Ford leadership reached out to Fred and asked him if he would take on a staff position. They explained that he would be responsible for determining what dynamics are important as Ford reduced the number of parts in a car. Fred also would essentially whisper in Frey's ear the things that needed to be changed within Ford engineering; he would arm Frey with the information Frey needed to combat the Ford engineers' linear thinking, negativity, and naysaying.

That Ford reached outside its ranks to find someone who could take on one of the most difficult problems it faced speaks volumes about both Ford and Fred. They offered him a staff, not a line, position, a role as an eminent advisor, as opposed to a manager of people. Presumably, they did this because they knew and appreciated that their recruit was an iconoclast, and not at all a manager. Everyone who dealt with Fred professionally and personally knew well that he had little skill for reading people. An example his son-in-law, Steve Richardson offers, is that Fred might be explaining why a rainbow has a prism of color; he might doodle, draw, and explain away; and all the while, he would not notice at all that he had lost his student early in the explanation. Steve says that Fred could not slow down enough to baby someone along on any topic.

Another "Fred-obliviousness" story has it that Fred once went to dinner at a friend's house. The friend had a 14-year-old daughter. As Fred was talking about cars, the friend's daughter became quite interested.

"Have you driven a car?" he asked.

"No," she said.

"Well, would you like to try? Come on out, and I'll show you how to drive."

Mid-dinner, he excused himself from the table and went outside with her to show her how to drive. Fred carefully taught her how to pilot the car around the circular section of driveway.

After determining that she knew how to drive, he came inside and sat down at the table to resume dinner.

"Where's our daughter?" the friend asked.

"Oh, she's outside driving around," Fred said, and continued eating. He considered driving a car to be a practical, real-world application of engineering that knew no bounds, such as age or gender. He had never considered that it was inappropriate to leave a 14-year-old outdoors, driving on her own.

On paper, the Ford Motor Company drew up the right job description for Fred's brilliant engineering skills and managerial deficits, and so . . . he agreed to move to Detroit and work for Ford. As empty-nesters—their youngest, Martha, was 18 and in college, and their oldest, John, was 27 and married—Fred and Martha would make the move from Dayton to Detroit. They put their beloved all-cement home on the market and moved to Bloomfield Hills, an upscale village of about 2,500 people, just outside Detroit. Because the largest and most profitable companies in the world had headquarters in the Detroit area, its Bloomfield Hills suburb featured impressive residences with sweeping lawns.

The Hoovens initially moved into a large, Colonial-style home that Fred had found on his own and purchased. Martha never saw the place before moving day. Parts of Fred's Dayton professional shop and basement workshop moved with them to their Lone Pine Court abode, loosely organized into a new basement workshop. The rhythm of his weekends, when he had visitors such as children and grandchildren, involved periodically going down into the workshop, doing something, and then coming back upstairs to mix, mingle, eat, socialize, or whatever else was called for on the main floor of the house. Then he went back down to his workshop. His grown children sometimes followed him down to chat and bond.

Not long after moving to Lone Pine Court, the Hoovens decided to commission the design and construction of a new home with an open floor plan. They purchased a lot at 910 Sunningdale Drive, and about a year later, moved into their capacious, newly built home.

Meanwhile, at Ford, Fred began what became a roughly decade-long struggle to infuse more engineering creativity into the company. His reputation had preceded him—people knew their new engineering guru was iconoclastic, unconventional, prolific, productive, and brilliant; they knew he spoke his mind; they knew he thought and acted independently. They had licensed a patent from him related to the rear axle system for passenger cars: patent number 2,753,190, filed in 1952 and granted in 1956, but assigned to Ford Motor Company. Grandson Mike says his father, John, believed it was the first patent Ford ever licensed. Whatever the case, at the time Ford certainly, at a minimum, did not often license inventions from others.

Fred served as executive engineer for advanced car products, envisioning improved engineering of the drive train and suspension systems of Ford autos. He worked on two main areas of engineering: the unibody car and front-wheel drive. Fred felt more passionate about front-wheel drive than unibody car. He ardently believed front-wheel drive represented the future of American automobile design and engineering. Front-wheel drive had been used extensively in Europe, mainly because of the smaller size of vehicles there and the preference for superior steerage and maneuverability around the smaller roads of Europe. In the US, Henry Ford had designed his vehicles as rear-wheel drive, and American automakers had followed suit. As such, only some American race cars had front-wheel drive.

Fred, however, had studied European auto design, and greatly preferred the front-wheel drive. As an engineer, he knew that designing the drive train to be at the front made much more sense. He thought it made little sense to have a heavy, expensive, inefficient system that connected the engine in front to the drive train in back when one could simply put the drive train up front. Plus, Fred felt automobiles should follow the model of the horse-and-carriage. Horses did not push carriages from the rear; horses pulled carriages from the front. Fred thought that, despite its unrivaled excellence in engineering and invention, America could occasionally learn something from the Europeans.

Fred had few allies at Ford advocating for this change to front-wheel drive. Alas, it seemed the entire company third largest in the world after GM and Chrysler believed that rear-wheel drive was the only way to design a car. At one point, a senior Ford accountant proposed that Ford could better reduce the cost of its vehicles by maintaining rear-wheel-drive engineering. Fred resented in the extreme an accountant making suggestions about car design and engineering, and he showed his resentment amply. In April 1959, he authored a five-page memo dripping with sarcasm. Sent to twelve of Ford's most senior people, it started:

> We have given very careful consideration to the letter received from Mr. R.C. Versey in which he has proposed that the cost of our vehicles be reduced by leaving the engine in its usual place at the front of the car, but driving through the rear wheels instead of through the front wheels as in conventional vehicles.

He proceeded to outline four engineering reasons why they could not do it, and one manufacturing reason. He sarcastically mentioned that the accountant's proposal would creatively enable a large tube to be put down the center of the car so as to keep occupants divided, and that the accountant's proposal would also raise the center of gravity for the car, which, although not very safe, would have the benefit of offering passengers a better view. He closed with:

> In conclusion, I would like to say that we have spent a great deal of time and effort in the evaluation of this proposal because of the high level from which it was sent to us. It is hoped, however, that in the future some simpler way can be found to reject such patently impracticable proposals before valuable engineering time is wasted in their evaluation.

Grandson Mike says that the accountants did not at all understand his sarcasm, and, indeed, they took seriously the heavily tongue-in-cheek points he made in the memo. They apparently tied themselves in knots trying to respond earnestly to the points Fred had made.

With his scruffy mop of gray hair and his plainspoken style, he was the most senior American auto executive to bang the drum (incessantly) for smaller, more fuel-efficient front-wheel-drive cars. Ford entertained their gadfly's ideas as interesting, but in resources and decisions, it basically turned a deaf ear and blind eye to his demands that Ford vehicles be front-wheel drive.

Ford turned Fred over to a different project, one that both parties felt would be more productive and satisfying. They gave him a new title, that of Technical Assistant to the Chief Engineer, and put him to work on advanced engineering for the 1960 Falcon, with his main task being to develop the unibody car for those models. Competitor Chrysler had coined the term "unibody" to describe the welding of the body and frame of the car into one lightweight cage. The unibody process and design saved time, money, and weight while enabling a higher-performing product. The Falcon would be the first Ford car to be unibody.

Fred did his job, and the 1960 Falcon fared very well. Today, it is a celebrated Ford car; it is the basis for the Mustang, first launched in 1964. In a similar unibody vein, Fred then worked on the 1961 Thunderbird, 1961 Fairlane, and 1965 Galaxie. The 1961 Thunderbird and Fairlane cars also fared very well in the market. With these successes in hand, Fred acquired a new title: special consultant to the general manager, and a member of the operating committee at Ford. They tasked him with determining how to redesign and reorganize Engineering and Research to be more innovative and creative. Fred issued a proposal, and Ford implemented it in 1967. He also wrote books, published under the auspices of Ford, such as *The Future of the Automobile in the United States* and, later, *The Potential for Automobile Weight Reduction Outlook as of 1975-1976*. With his writing, his vision, his designs, and his successes, he and his ideas and products represented the future of the industry.

Although he led projects that enjoyed commercial success, Fred did not enjoy his time at Ford. Ford executives might not have enjoyed his presence there, either. Although they valued his contributions, they reportedly found him to be cantankerous and difficult, a skunk at the garden party, a fly in the ointment, a necessary evil. They knew he was too brilliant not to put in charge of their most important and difficult projects, and they believed that as long as they surrounded his insights and designs with accounting, marketing, and operating people, they could harness his brilliance. But Ford never let Fred practice his craft openly, with candor, and with resources behind him. They required that he operate under the watchful eyes of accountants, marketing executives, and others. Fred found this modus operandi to be excruciating. He could not entertain the misguided policies of Ford executives.

As evidenced by the memo Fred wrote excoriating company accountants, Fred thought the Ford accountants and finance people were especially dim, bordering on ineducable. "He had few kind words for the accountants who oversaw everything he tried to do at Ford," John Collier, a colleague from Fred's subsequent days at Dartmouth, says. Fred reportedly struggled to accept and tolerate the fact that the accountants' very short-term and narrow views could determine whether a car was better for the consumer and better for the world. He felt the numbers people stifled greatness, advancements, and progress, and that they instead came down, time and time again, on the side of design and engineering mediocrity. Plainly frustrated with the bureaucratic mindset, he walked the halls of Ford as an irascible, visionary, brilliant man, and he advised executives throughout the corporation and, indeed, the industry on advanced automotive engineering. Fred knew the company was harnessing and containing him, rather than deeply valuing his vision for an improved line of automobiles for contemporary Americans.

Executives tolerated Fred because they knew he would be powerful in the hands of the competitors. They had to make use of him.

They could not bring themselves to part ways with him. As president of Ford, Lee Iacocca (later CEO of Chrysler) noted that Fred's nature was starkly different from that of the system in which he played. Iacocca had a salesy style, which he employed in bringing two companies—both Ford and Chrysler—to great fortunes. This revered, senior-most executive in the industry considered Fred to be brilliant and difficult. According to grandson Mike, Iacocca "kept trying to fire" Fred. Evidently, he could not quite bring himself to do it; he evidently consistently found Fred to be too valuable. (To wit, Fred's development of the Falcon became the basis for one of Iacocca's great legacies, the development of the Mustang.) Iacocca considered Fred a necessary armament in Ford's attempts to build better cars and gain market share. He appreciated the role Fred played; he knew Fred wanted to advance the industry, to improve automobiles in the US for consumers, and that he never cared a bit about his own standing or enrichment. Fred Hooven wanted automotive engineering to be the best it could be, especially in the form of lightweight, efficient vehicles. Here's what Iacocca later admitted:

> The thing I remember most about Fred is that he said future cars would not be built the way cars were built then. Front-wheel-drive was the way of the future and rear-wheel-drive was antiquated. He would say, "It's silly to design cars the way we do. Why not put a power pack up front just like a horse? A horse will pull anything. Behind it you could put a fire truck, a station wagon, two people, four people, six people limousines."
>
> And of course it turned out that way, the way Fred said it would. We do have front-wheel-drive minivans today that were a glint in his eye then because he said that is the way to do efficient packaging.

In 1967, as consultant to Ford's General Manager and member of its Operating Committee, Fred determined he wanted to get out of corporate life. His son Mike says he developed an ulcer, probably because, for years, he had been in a role contrary to his skills and preferences. He was a doer, a maker, not an advisor. He risked wilting amid the bureaucracy. He wanted more influence and power on matters related to engineering. He retired.

Gordon Buehrig, the brilliant designer of the Cord, certainly enjoyed working with Fred. In his interview with an automotive magazine, he said:

> . . . Fred Hooven was a real genius in engineering. But not only that. I've always felt that I had a little bit of talent in my field, but this guy had talent in so many fields. He could have been a concert pianist; he was just wonderful on the piano, and had a whole repertoire of classical music in his head which he could sit down and play. He would get frustrated at Ford Motor just as I did, and if he was at a meeting and got frustrated he would just walk out and drive over to our house, and sit down at the piano and play for a while; that would quiet him down. Then he could—well, this is not a big problem any more with computers, but it was amazing to me—he could do the cube root of, say, a five-digit number in his head. He was an astronomer too. He just knew about everything; I've never known anyone who had so much knowledge in his head as this man did. He was a brilliant expert in everything—and a very kind person along with it.

As a postscript to the front-wheel-drive saga, Ford ended up licensing the patent it had in-licensed from Fred to General Motors. The result of GM gaining license to the patent became, in a nutshell, GM's new front-wheel drive Oldsmobile Toronado and Cadillac Eldorado automobiles. These cars sold incredibly well, became renowned within the automotive industry and general public, and fundamentally changed US automaking. Fred had had nothing to do with them, but his patent, licensed to Ford and then, in turn, to GM, had everything to do with them. The reason for GM's wise and productive use of the patent was, in Fred's estimation, the culture established by the GM engineering head, Charles Kettering, as opposed to the Ford culture. Fred had always admired Kettering, and considered him a friend. Boss Kettering, as he was called, was a prolific inventor-engineer and head of research at General Motors from 1920 to 1947. He earned 186 patents, including the patent for the electrical starting motor. Both Fred and the GM boss possessed a pathological drive to find and solve problems by using mechanical and electrical engineering, and both did so at a dizzying pace and with tremendous success. To wit, Kettering famously said, "I notice the harder I work, the luckier I get," and "Failures are finger posts on the road to achievement."

Kettering and his family admired Fred, adding him to the Kettering Foundation board. Boss Kettering, in fact, said that during the time Fred worked at Ford, he was "a Ford trade secret." True, and yet perhaps it's more accurate to say that, in playing a main role in changing the mechanics and construction of American cars, Fred might even have been "an American trade secret."

III

Fred loved reading *Scientific American*, a popular science magazine that—having been founded in 1845—is the oldest monthly magazine in the country. In addition to being published in the magazine, Fred loved communicating with Martin Gardiner, its longtime "Puzzle Master." Fred often pointed out an alternative way to solve a complex mathematical problem. He communicated so often with Gardiner that they became good friends.

In 1967, *Scientific American* held the First International Paper Airplane Contest. In publicizing the contest, the magazine article stated that the latest in supersonic plane designs bore a striking similarity to the simplest paper airplanes. The magazine advertised its contest broadly, including in *The New Yorker* magazine. Fred entered the contest in the flagship "duration aloft" category, which had nearly 11,000 entries. He designed a laughably, shockingly simple plane, the design based on wind tunnel data that the Wright brothers had used to design their flyers. Here's what the man in charge of advanced engineering for a top-three largest auto company in the world did. He took a half- piece of paper (it might have been regular paper or possibly lighter-weight tracing paper) and folded half of it twice and then over on itself to create a leading edge. He cambered the leading edge slightly and put a slight crease in the tailing edge to give it lift. It looked and performed just like the simple Wright brothers' single-wing. Those who saw it said the contraption looked like it could never work.

The key to the plane, of course, was in its simplicity of design and physics. The thrower of the plane did not use any arm at all. Indeed, one did not throw it at all; one simply set it free. With hand in air, holding the plane, a thrower let go of the plane and off it went at the speed of a slow walk, for 1 . . . 2 . . . 3 . . . 4 . . . 5 . . . 6 . . . 7 . . . 8 . . . 9 . . . 10 . . . seconds, plus a fraction. At 10.2 seconds, Fred's plane had the longest glide time among nearly 11,000 entries. When interviewed, Fred told a newspaper that he learned about planes from Orville Wright. "Mr. Wright loved toys almost as much as he loved airplanes. We were good friends."

Fred's victory stemmed not from luck or chance, but, rather, long-time, skilled study. He had studied the Wright brothers and had spent a career in aviation engineering. He knew the math and physics of wind. He understood simplicity and innovation. The magazine gave Fred a trophy of a hand holding a paper airplane and issued a press release about his plane's design. A number of papers ran articles about Fred's win, including the *Detroit Free Press*, the main paper Fred read. Next to the article, the paper ran an article about Monroe Hickson, who had recently landed on the FBI's Ten Most Wanted Fugitives list. Hickson had recently escaped from a South Carolina prison, where he was serving four life sentences for murder. However, the *Free Press* mistakenly put Hickson's image as Fred Hooven's, and put Fred's image as that of Monroe Hickson. The caption named Fred as one of the most wanted fugitives in the country. Fred got such a kick out of the transposition of pictures and the caption showing him as a fugitive that he framed the article and hung it on his study wall.

|||

Fred loved to draw and, as with most activities he undertook, he did it very well. (The drawings for his patent applications are works of art.) He especially enjoyed drawing trains and cars, and as board director of YSI, he drew a lot of them. YSI was the originally named Yellow Springs Instrument Company that Fred had helped found. Board members adored him, but knew he could not tolerate ambivalence, indecision, or mundane discussions; all three of those traits in a single meeting could nearly drive him mad. To stave off sleep during boring discussions of what he considered trivial matters, he doodled . . . obsessively and compulsively . . . throughout most board meetings. Directors reportedly thought this to be humorous, as they knew they needed to pay this small price in order to access his incisive insights on key decisions. When he retired, the company directors compiled a booklet, "Odyssey of a Bored Director," that featured some of his sketches.

One director, Pete Stern, wrote:

> Dear Fred, Your engineering drawing is an art form; so has been your ability to come wittily to the point at YSI directors' meetings. It's as simple as this: meetings you attended were marked with verge and a feeling of expectancy. I will greatly miss you.

Others wrote:

The fostering of humanism in an organization and of excellence in innovation, technical or otherwise, are widely recognized and treasured Hooven trademarks.

[You are] a challenger of erroneous assumption, a cutter-through of red tape, an encourager of risk commensurate with gain, a spurner of obfuscation, and a real pleasure to serve with as director.

Your outstanding technical insight with a ready appreciation for the other qualities of a person will always be remembered."

and

"I'll remember your impatience with the trivial, and your quick and eager joy in the beauty of good science."

✿

Beloved Professor

Fred had long thought about a role as a professor, and had harbored a desire to teach at MIT, his alma mater. While working unhappily at Ford, he spoke with MIT administrators and professors. They told him they wanted to hire him, but had a policy of requiring a PhD for a tenured professor. This decision, by what Fred called "the wrinklies at MIT," frustrated him, but not for long, because another institution wanted his time and talents.

Two Dartmouth professors, Dr. Robert C. Dean, Jr. and Dean Myron Tribus, knew of Fred and hoped to recruit him to faculty. As head of Dartmouth's Thayer School of Engineering, Dean Tribus had recently led the faculty in developing a new curriculum based on hands-on engineering design and entrepreneurship. He would go on, soon after bringing Fred aboard, to serve as Assistant Secretary of Commerce for Science and Technology in the Nixon Administration, a senior vice president for research and engineering at Xerox Corp., and then the head of the Center for Advanced Engineering Study at MIT.

Dean Tribus and Dr. Dean asked Fred if he would consider coming aboard as adjunct professor of engineering. He did not take long to consider their offer—the opportunity to teach with colleagues he respected at a school that ranked among the top in its field thrilled Fred. He and Martha moved from Michigan to Norwich, Vermont, a couple of miles from the Dartmouth campus. Norwich is a tiny, New England town with a general store, a town square, Colonial-style houses, the Connecticut River, adorable farmhouses, and well-to-do residents. The Norwich climate is colder in winter than Bloomfield and Dayton, but more temperate in summer than both places.

Fred loved it! As usual, he demonstrated his pioneering ways by commissioning the construction of a new home—not just any home, but a deck home. A deck home is a modern style of house that came of age in the 1960s, and Fred's had cantilevered beams, cathedral ceilings, and vast windows that allowed terrific views of the White Mountains. Fred had his home prefabricated elsewhere and the parts shipped to Vermont for construction. His Elm Street abode had a machine shop in the basement in which sat desks, bins full of supplies, a drill press, a belt sander, a couple of grinders, and his favorite, the Atlas metalworking lathe.

At work, Fred quickly became a beloved professor, easily and graciously filling his role to advise and aid Thayer's brightest students on special projects. He explained deep, difficult-to-grasp engineering topics in a remarkably relaxed and informal manner. In his uniform of a tweed coat and occasional tie, and with his unruly gray hair, he developed more of a fireside chat feel in his advisory seminars than a lecture-type feel, and when more discussion was needed, he invited students to have dinner with Martha and him. He plainly loved dealing with Thayer's aspiring engineers and faculty to solve engineering problems.

Fred told his colleagues he wanted students who wanted to learn for learning's sake. He explained that he would love to set up a social experiment. In the experiment, on the first day of class, professors would tell students they all will receive an "A" grade whether or not they come to class or even do any work. On day two, the professors will see who is actually interested in learning—the ones who show up on day two are the ones who must truly have interest in learning.

Thayer colleagues admired not only how energetically Fred paid heed to his students' questions and problems, but also that he possessed what they thought to be a rare combination of skills. Fred was much more than an academic engineer. They saw that he machined items with great skill and precision; he constantly designed and ran experiments; and he performed analytical work on experiments as well as anyone they had ever seen. This practical, experimental, and analytical combination stood out in the academic environment of Thayer.

Not only did Fred's Dartmouth colleagues notice his unusual combination of skills, but so, too, did some of the best business minds in the country. In 1983, the business management guru, Tom Peters, wrote about Fred's rare combination of building, testing, and analyzing. In his renowned bestselling book, *In Search of Excellence*, Peters wrote:

> When 'touch it,' 'taste it,' 'smell it,' become the watch-words, the results are most often extraordinary. Equally extraordinary are the lengths to which people will go to avoid the test-it experience. Fred Hooven, protégé of Orville Wright, holder of thirty-eight major patents, and senior engineering faculty member at Dartmouth, describes a ludicrous, yet all-too-typical case: "I can think of three instances in my career in which my client was making no progress on a complicated mechanical problem, and I insisted that the engineers and the technicians (model builders) be put in the same room. In each case the solution came rapidly. One objection I remember being offered was that if we put the engineers in the same room with the shop it would get the drawings dirty." Hooven adds, in support of the overall point, "The engineer must have immediate and informal access to whatever facilities he needs to put his ideas into practice . . . It costs more to make drawings of a piece than to make the piece, and the drawing is only one-way communication, so that when the engineer gets his piece back he has probably forgotten why he wanted it, and will find out that it doesn't work because he made a mistake in the drawings, or that it needs a small change in some respect, which too often takes another four months to make right."

Another area of excellence at Thayer involved one of Fred's great interests, paper airplanes. Fred occasionally replicated his winning paper airplane in the Thayer hallways. He stood in a common area and simply let it go—no active "throw" involved. Some of the nation's top would-be engineers marveled as they watched the folded paper stay aloft for more than his contest-winning ten seconds. Not only did the duration aloft impress them, but equally entertaining was the laughable simplicity of Fred's design. Grandson Mike recalls, "You could just hold the plane over your head and release it, then walk alongside it for the length of the room." The Dartmouth campus newspaper ran an article for an event, held at the school's squash courts, where Fred would pay ten dollars to whoever could beat him in a paper plane duration-aloft contest.

Simplicity was the brilliant aspect of Fred's teaching style. He knew where he was going with each seminar and what he would teach throughout a semester, but he managed student discussion such that they led him on that path. He encouraged them to ask any and every question about anything they wanted, and then somehow (people say it was a marvel to watch but impossible to describe) figured out how to bring them around to whatever point he planned to make during the class. He never worked from notes or books, but, rather, drew, wrote, and scribbled on the chalkboard. On rare occasions, he used an overhead projector, perhaps to show a flight path or aviation mathematics. He preferred, however, to use his words to paint mental pictures of advanced, unconventional products and components.

It's interesting that Fred, by every account, taught students extremely well at Dartmouth, despite the fact that he had reportedly always been unable to "read a room," and despite the fact that he often did not know when he was going above someone's head. Perhaps the high intelligence level of students at Thayer and Fred's relaxed joy in teaching with no need to manage, as well as his age and maturity, intersected to establish the right set-up for him. Perhaps at Thayer, he could relax enough to notice where he needed to slow down and advise, mentor, and teach other very bright people. But this is speculation.

John Collier, whom Fred mentored on his bachelor's, master's, and doctoral engineering projects, and who would later become dean of engineering, recalls that Fred attended many lectures delivered by engineering professors and graduate students, and he livened up those sessions. "Fred would be sitting in row two or three and, occasionally, halfway through the session, I'd see his head drop to his lap," Collier says. "I'd figure, *Fred's fallen asleep.* Some people in the audience would ask some minor questions, with Fred's head still drooped nearly into his lap. Then, out of nowhere, Fred would lift his head, raise his hand, and ask a cutting, deeply thought-out question that would penetrate to the core of this person's research. If the research was not sound, he would just slice it in half."

When Fred went to master's- or doctoral-degree student presentations, he thought at a level so profound that people sometimes did not know where he was coming from, but as soon as they figured it out, they determined his question was cutting, incisive, and important. Although he lacked a master's or PhD degree, Fred had the full respect of faculty and students at every level. Collier recalls, "One of the things I learned when I took Fred's course was that you have to fail. He used to say: 'Fail early and often because you learn from every failure. You don't learn if you don't fail, and you're not learning if you're not guided to understand those failures.'"

With Collier, who eventually became dean of Thayer, Fred enjoyed a marvelously collegial relationship. Collier later said of Fred, "Of all the people I've met, he is one of the very few true geniuses. I have met lots of Dartmouth faculty, and he's right at the top. But also, as a warm, friendly, supportive, encouraging human being, he is again at the top." Dean Collier describes how Fred regularly came in and asked him what he thought about some idea or another and then quips, "It was like Patton asking one of his tank commanders, 'Do you mind if I run this by you?'" He further describes Fred's impact. "Fred was a wise man—one of those fully developed human beings whom we are sometimes privileged to encounter during a lifetime . . . his presence was like an oracle. People came to him and asked questions all the time . . . He was like Warren Buffett at Berkshire Hathaway when you have a problem, find Fred!"

After eight years as adjunct professor, in 1975, Fred became a full professor (no longer adjunct) and continued apace, working on various engineering projects, papers, and inventions. In 1978, as part of 75th anniversary of Wright brothers' first powered flight, he conducted a computer analysis of their first plane. Collier recalls that Fred for a while could not figure out the math and physics of the Wright brothers' flights. He could not make sense of the wind tunnel information and other information he had about their planes and flights. Some aspects of the Wright brothers' flight physics perplexed him. Actually, they were driving him crazy.

Then, one day, Fred came bounding into Collier's office. "I got it!" he exclaimed, and proceeded to explain to Collier that the Wrights had designed their plane to be inherently unstable. Why? Because the Wright brothers were longtime bicycle makers, and the bicycle is inherently unstable. It relies on operator skill and balance. The Wrights brought that same "operator focus" to the plane. "The plane was never supposed to be stable!" Fred exclaimed with glee. Fred had not, as he had long thought, been incapable of parsing the math and physics of the plane and its flights, nor had the Wrights been wrong to design the plane as they had—they had made that personal choice based on their experience and worldview.

Fred excitedly composed a 1978 article that *Scientific American* published, "The Wright brothers' Flight-Control System." He started the article itself with this:

> Wilbur and Orville Wright invented the controllable airplane. Until they first flew in public in 1908 it was believed a powered aircraft would be similar in its behavior to an airship, a stable vehicle that could be steered right and left by a rudder and up and down by a horizontal rudder or elevator. One could expect to mount such a craft and fly it without previously acquired skills, and that was what was invariably attempted until the Wrights showed how it should be done.
>
> In contrast, the Wrights, who were builders of bicycles. conceived the airplane from the beginning as being a vehicle that like the bicycle depended on its operator not only for its direction but also for its equilibrium. It therefore seemed perfectly natural to them that before one could hope to successfully operate a powered aircraft one needed to develop both the aircraft and the skills necessary for operating it.

Fred also worked during his Dartmouth years on a prosthetic orthopedic bone replacement, lightweight auto, computerization of medical diagnoses, and a digital music synthesizer. That last item, the synthesizer, is a fun development that illustrates Fred's creativity, intelligence, and collegiality. In the early 1970s, composer and Dartmouth music professor Jon Appleton wanted to combine a mini-computer with a music synthesizer. The main commercial synthesizer available to Appleton at the time was the Moog synthesizer, which linked a piano keyboard to an analog computer. It had no memory, and Appleton wanted memory. He turned to the engineering prowess at Thayer.

He worked with a research professor, an undergrad engineering student, and Fred to build the prototype. Together, in a project involving hardware and software labs and the music department, the dream team created the Dartmouth Digital Synthesizer. Started in 1973 and completed in 1975, it was, according to the many websites and forums dedicated to music synthesizers, the pioneering prototype hardware and software system for all "digital non-linear synthesis, polyphonic sampling, magnetic (hard-disk) recording and sequencing systems technology." Appleton later said the prototype the team built was precisely what he had wanted. "It did so many things, and the software was so beautifully integrated."

The LinkedIn Learning (formerly Lynda.com) website says that it was developed by Appleton and Hooven and "powered by a 16-bit mini computer processor controlling 8-bit additive FM and timbre frame synthesizer voices. There's a green screen, or CRT monitor, which is also used to enter and edit sounds, music events, and computer files. These were all originally stored on 5.2 floppy discs, and later on, hard disc drives."

The team's research professor and engineering student left Dartmouth (after graduating) and co-founded New England Digital to commercialize the synthesizer. New England Digital branded the world's first digital synthesizer as the Synclavier Synthesizer System. A second generation followed, and then a third, with the brand name evolving into the Synclavier Digital Audio System, or sometimes, Tapeless Studio. The company sold its synthesizers at a price range of $75,000 to $500,000, and the industry absolutely loved them. They considered the master-digital-sampler-and-synthesizer workstation to be first-in-kind and far ahead of its time. The high price tag made it feasible for only the most successful musicians, commercial studios, and sound designers—the instrument of choice for many renowned musicians, composers, and others. A short list includes The Cure, New Order, George Duke, Pat Metheny, Sting, Stevie Wonder, Paul Simon, Foreigner, Michael Jackson, Depeche Mode, Genesis, The

Cars, Soft Cell, Frank Zappa, and film composer Alan Silvestri. The Synclavier was used to score the film *Apocalypse Now*.

Numerous websites have special sections dedicated to Synclavier parts. The user guide alone for one model of Synclavier is 196 pages and seven chapters. It's said that there were so many upgrades and improvements to each model that finding user guides for each specific sub-model of the system is an exercise in frustration. Today, the Synclavier is considered the Rolls-Royce of music synthesizers, and Fred played a substantial role in its development.

III

The last patent Fred submitted, coinvented, and assigned to the Thayer School was for a connector assembly for coupling optical fibers together. A fiber optic coupler is a device that can distribute the optical signal from one fiber among two or more fibers or combine the optical signal from two or more fibers into a single fiber. Fiber optic couplers are needed for monitoring signal quality and for telecommunication systems such as ring architectures that require more than simple, point-to-point connections. Although Fred had never previously worked with fiber optics, he designed a fiber optic coupler and submitted the patent. The summary for the European Union patent states:

> The invention provides a connector assembly for optically coupling a pair of optical fibers together. The connector assembly includes a first lens means adapted to be coupled to one of the optical fibers and a second lens means adapted to be coupled to the other one of the optical fibers. A first and second lens means are axially aligned and spaced with respect to each other for directing light rays paraxially (i.e., a collimated beam) there between to eliminate the light loss effects due to fiber spacing and focusing the light provided by one optical fiber onto the other optical fiber.

John Collier says of this invention, "Fred came to me and said, 'You know, optical fibers are going to be big in communications, and the biggest problem with optical fibers is making the connectors. Let's figure out how to make connectors. So Fred got optical fiber and built sophisticated lenses, talked with an optics company, and made some lenses. These were some of the very first optical connectors between optical fibers, and they were made in our lab under the guidance and tutelage of Fred."

Grandson Mike reminisces about a conversation related to fiber optics during which his grandfather described a cable connection problem he had worked on for a big company. Fred explained that the company had to use enormous machines to provide enough force to crimp the big cables together. The company had trouble transporting the large machines to remote areas. "Grandad said that he had asked, 'Why don't you use an explosive within a confined chamber over the connectors to blow the cables together?'" Mike says he wonders whether Fred, had he not been approaching old age at that point, would have done for fiber optics what he had done for avionics and automotive mechanics.

III

For his deep and broad capabilities in engineering—capabilities that showed themselves to extend, across the expanse of a lifetime, from mechanical to electrical, from automotive to medical, from analytical to practical, and from theoretical to productive, the elite National Academy of Engineering admitted him into its ranks (1979). That same year, Fred wrote "Invention and the Art of Engineering," an intense and sweeping paper describing the value of engineering to societies from prehistoric times to the contemporary era. Although it's unclear where the paper was published, it might have related to his induction into NAE.

✿

Grandfather

In Dayton, before moving to Bloomfield Hills, Fred and Martha became grandparents. Oldest son, John, married in 1952, and he and his wife, Betty Ann, had their first child a few years later. The Hooven grandchildren loved visiting their grandparents' Bloomfield Hills home. Martha thoroughly fulfilled the role of kind, interested, nurturing, loving, and supportive grandmother. Fred maintained his pattern of periodically going downstairs into his basement workshop. His family visitors observed him traipse down the basement stairs, presumably to do one thing or other with a machine, come back up, interact with them for a bit, and then traipse back down the stairs to repeat the cycle. His grandchildren often followed him down the stairs, and in his workshop, they talked with him about engineering and other matters. They worked with him on small projects such as, for instance, making and then threading a screw—the basic activities of adolescents beginning to learn how to make things.

Upstairs, Fred interacted with friends and family by probing, questioning, being insatiably curious, and then challenging. He did this with joy, a smile, and enthusiasm, rather than with criticism, and the effect on family members was to inspire, rather than intimidate; everyone loved going to Fred and Martha's house. Family members and others say that virtually no one in the family or among his non-colleague friends understood the profundity of his accomplishments in science, industry, and academia. Fred rarely spoke of them.

"Whenever we went to visit my grandparents, it was like going to Disney World. There were always such cool things that he developed since the last trip," grandson Mike says. "When Grandad entered the room, the energy level went up exponentially, as did the activity, the engagement, and the fun." Mike also recalls that his grandfather "always had a twinkle in his eye and laughed a lot." To him, Grandad acted like a big kid, enjoying playing games, making toys, doing puzzles, working out math problems, and such activities that stimulated the brain and were tinged with intellectual adventurousness.

He tells a story about Fred's math prowess. While in the car on one of his family's many trips to Bloomfield Hills from their home on the East Coast, Mike wrote down the most complicated math formula he could think of.

"I took three or four lines of paper to write the full equation. I spent time calculating and recalculating the correct answer, which I kept handy. Upon arriving in Bloomfield Hills, I showed the equation to Grandad to see if he could give me the correct answer. Handheld calculators weren't really around. Grandad carefully read the equation, and right away (almost before he reached the end of the equation) gave me the correct answer."

This stump-Grandad game became a fun tradition every time adolescent Mike visited his grandparents in Michigan.

"I was never able to stump him," Mike says.

In his spare time, Fred continued to pursue myriad interests. His passions for photography and trains intersected with each other in 1974, the year in which he built a model train and decided he wanted to take a photograph of it in front of Vermont's White River Junction train station. He loved the challenge of perspective—that is, making the enormous size differential between a roughly six-inch-high model train and a roughly thirty-foot-high, real-life train station work seamlessly in a picture. The perspective problem involved math, geometry, and engineering. Fred rebuilt his 35mm camera lens to enable extreme field depth, and he used the newly engineered camera to produce a like-real photograph of the train in front of the train station. *Model Railroading* magazine put the image on its February cover, and any reader who did not read the information about the making of the cover photo would never have known that the photo had an extreme scale adaption involving a miniature model train in front of a real, full-size train station. (Later, Fred made further adaptations to his 35mm camera to enable remotely activation. This ensured he could capture dozens of stunning photos of hummingbirds feeding at a bird feeder in his yard.)

In other areas, Fred kept building ever more. He built paper and balsa wood airplanes for his grandchildren. Sometimes he added carbon dioxide cartridges to the planes, an engineering advance that his erstwhile Thayer recruiter, Myron Tribus, wrote were "perhaps the first jet-propelled model airplanes."

Another time, he invented something for Martha who, in her old age, developed a back problem that forced her to lie prostrate all day for several weeks. She told her husband she would love, at a minimum, to be able to read books during her convalescence. Fred agreed that lying down with no ability to do anything was not good and that she needed at least to be able to read. He quickly developed for her special glasses with lenses consisting of 90-degree prisms. That way, she could read the book while facing the ceiling and holding the book vertically at her waist, which is precisely what she did.

Fred also seemed to become more politically passionate as he aged. Of his early, strong, but less-often-expressed political views,

Fred's son Mike recalls, "I remember once in the early 1950s, my father and grandmother [Fred's mother, Anna] were having a political argument in our living room. My grandmother liked Robert Taft, a senator from Ohio"—he unsuccessfully ran for Republican presidential nomination three times—"when Adlai Stevenson was the Democratic nominee. Dad strongly disliked Taft. I had never heard such things coming from my father or grandmother, but they were really going at it. They were worked up."

Throughout his career, Fred had spent his energy mainly discussing engineering, problem solving, building things, and education, much more than discussing politics, but that changed a bit as he aged. He served on the board of the Charles F. Kettering Foundation, which his friend Charles Kettering had founded to advance the notion of scientific research for the benefit of humanity. The Foundation over time adapted its mission to be more about promoting the role of citizens in democracy. This aligned with Fred's views on educational institutions as vital promoters of democracy.

As the 1980s rolled around, Fred told people that he hated President Ronald Reagan, except . . . he could not say he "hated" him because Martha did not allow the H-word in her home. "We do not hate," she was known to say. "We strongly dislike." Chastened, Fred said he strongly disliked Reagan.

✿

Legacy

In the late 1970s or early 1980s, Fred developed a heart problem. It went undiagnosed, so to deal with the symptoms, he began to self-medicate. Some combination of the heart condition, the self-medication side effects, and depression from not knowing what was going on with his body and mind caused him to think he was losing his mind. For Fred, for whom the life of the mind was everything, this caused profound distress. On February 5, 1985, at age 80, he chose to end his life.

As his children dealt with his sudden choice to end his life, many newspapers carried his obituary and celebrated his extraordinarily productive life. *The Boston Globe* wrote, "Mr. Hooven's inventions included radio direction finders and short-range radar for World War II bombers, ignition and landing systems for other aircraft, front-wheel drive and suspension systems for General Motors Corp. and Ford Motor Company automobiles, phototypesetters, computers and the first successful heart-lung machine."

In the National Academy of Engineering Memorial Tributes edition, Myron Tribus, the dean of Dartmouth's engineering school, wrote:

> Fred Hooven enjoyed engineering. In fact, he enjoyed everything he did. Hooven was interested in both model railroads and photography. He rebuilt a lens for a 35mm camera to provide

extreme field depth and used it to produce a photograph of a model locomotive in front of the train station at White River Junction that was so skillfully done that it looked like an actual locomotive. (The photograph was even used on the cover of *Model Railroading* magazine.) . . .

He liked to study whatever was new. He conversed intelligently with others about special relativity and quantum mechanics. While in his seventies, Fred Hooven continued his innovative work in the areas of prosthetic orthopedic bone replacements, music synthesizers, lightweight autos, and computerized medical diagnoses.

Fred was truly a classical engineer. He viewed the world's problems in terms of their potential solutions. His impact on students and associates was extraordinary. Fred could stretch the reach of others: He could make them broaden their horizons in terms of the problems they tackled and the ways in which they approached them. Fred Hooven was truly an inspiring teacher, colleague, and friend.

As the Thayer dean, John Collier set up the Frederick J. Hooven Scholarship Fund. The school posthumously granted Fred the annual Robert Fletcher Award recognizing "distinguished achievement and service in the highest tradition of the School." The school also established a "Frederick Hooven" level of philanthropic gift that offers donors at that level special access to school events and activities.

Frederick Allan Goerner, author of *The Search for Amelia Earhart*, wrote a letter to Fred's son John, stating, "The only other person I can recall who matched Frederick The Great was the late Fleet Admiral Chester Nimitz, who always displayed the same qualities: Genius, sensitivity, generosity, kindness, warmth of spirit, joy of life."

Martha Kennedy Hooven remained in Norwich until she passed away in 1997. The YSI stock from years earlier sustained her financially during the twelve years that she lived without her beloved husband.

The ten Hooven grandchildren have shown the creative energy and intelligence of their grandfather. Grandson Mike commissioned and wrote the foreword to this book in order to honor his grandfather's inspiration. After Fred died, Mike gathered his grandfather's entire workshop into a U-Haul trailer and brought it back to his Cincinnati home. In Mike's basement are the lathes, drill presses, benches, drawers, storage compartments, and more. Mike used his grandfather's machines to develop multiple medical devices and businesses, including the device that gave rise to AtriCure. AtriCure has grown into a $3 billion public company, saving and enhancing many thousands of lives. Mike then founded another medical device company, Enable Injections, which is enjoying similar success; it makes a disposable mobile device that replaces intravenous infusions and allows patients to self-inject biologic drugs with the touch of a button. Once again, Mike used the tools and machines from his grandfather's workshop to create the initial prototypes.

And then there is Fred's daughter, Martha, and her husband, Steve Richardson. Fred had told Steve that Steve could use his jigsaw to make puzzles for the company Steve wanted to buy. Steve subsequently renamed the company Stave Puzzles, put a potbelly stove in a basement workshop in Norwich, and used the jigsaw to produce the puzzles. Soon, the company became successful enough that his wife, Martha, could quit her job to join Stave. Not surprisingly, Fred became moderately involved with the company. For instance, at one point, he found out that Steve was struggling with an aspect of puzzle production using Fred's jigsaw. Fred quickly reconfigured it and, according to Steve, "solved a big headache for the company." Steve says the upgrade Fred quickly and intuitively made to the jigsaw was so smart as to become a Stave trade secret. Steve put a black cloth over the upgraded portion of the machine so it would never show up in pictures, where competitors could copy it.

Today, Stave produces custom-made, hand-cut puzzles. It has prospered; many puzzle masters consider it to be the world's greatest jigsaw puzzle company. Regular customers of Stave include the Lilly family (of Eli Lilly), British royalty, and Bill and Melinda Gates, all of whom have visited Stave in Norwich, Vermont. The Richardsons visited the Gates home twice (before the couple's divorce) to create custom puzzles for the family, and it was on one of those visits that Amazon's Jeff Bezos became a customer, too. Suffice it to say, the multi-million-dollar business produces some of the most creative puzzles in the world. The company used Fred's jigsaw for puzzle production for 45 years, until 2020, when it became too old and had to be replaced.

When asked for words that describe Fred, family, friends, and colleagues used words like engaged, curious, complicated, wise, gifted, fun, friendly, outgoing, encouraging, intelligent, clever, skilled, determined, thoughtful, and kind. He surely personified his own unique combination of those traits, but this book is about one that he embodied most of all—inventiveness. Frederick J. Hooven was brilliantly inventive in the combined quality, quantity, and diversity of his inventions. Only a brilliant man, probably a genius, can go from inventing and/or building a radio direction finder to a bombsight to a crystal phonograph pick-up to a phototypesetter to a heart-lung machine to a digital music synthesizer to a fiber optic coupler. The man invented and engineered across wholly dissimilar categories. When considering this inventiveness, which is preciously rare for its combined depth, breadth and impact, one can say that Frederick J. Hooven, 1905-1985, belongs among Ohio's top inventors and engineers of all time.

Fred as a young man. He had insatiable curiosity from a young age, and he loved engineering and designing things. As a Dayton resident, he befriended the nearby Wright brothers. For college, he attended MIT, where he majored in engineering.

In his early 30s, Fred designed and developed the SHORAN bombing computer. It consisted of a radio direction finder (aka, radio compass), a tuning meter, a loop antenna, a control box, and a mounting. He later said of his invention, "It was completely my idea."

The radio direction finder, aka radio compass, aka Hooven RDF, that Fred invented.

Tuning meter that Fred invented.

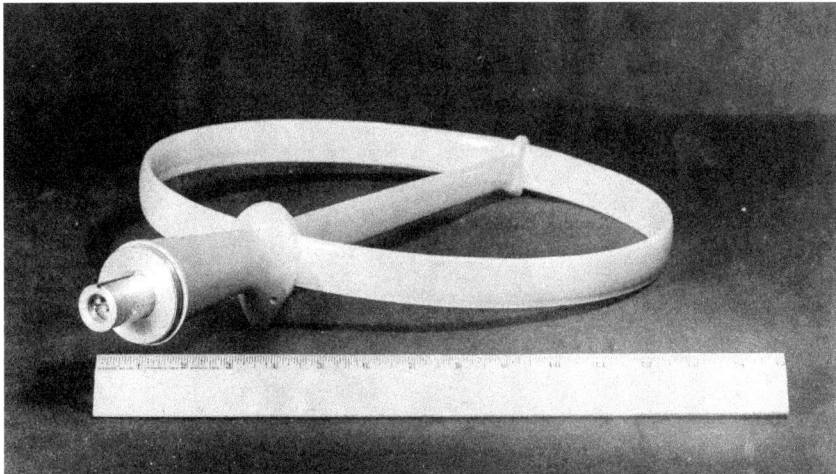

Loop antenna for the radio compass.

Control box for the SHORAN.

Mounting for the radio compass.

Six Airmen from the USAF 310th Bombardment Group, a highly decorated specialist group that participated in the Euro-African-Middle East campaign of WWII. They used the SHORAN system that Fred invented.

The USAF 310th Bombardment Group flying through the clouds.

This shows the target for a bomb dropped by the USAF 310th and where the bomb actually hit—a near bull's-eye! SHORAN was much more accurate than the Norden Bombsight system.

The Dayton Daily News wrote about Fred's invention being commissioned into the Air Force Museum.

A DC-3 plane. Fred said in an interview, "[The automatic direction finder that I invented] completely dominated the scene for that kind of device for a time roughly corresponding to the life of the DC-3. It made it routine to cross the ocean, where it had been an adventure before." (Photo courtesy Towpilot, Creative Commons Attribution-Share Alike 3.0 Unported license.)

Fred built this car with his son, Peter. Fred is in passenger seat. Driver is Harley Copp, lead engineer of numerous Ford car models. Ford's lead designer, Gordon Buehrig, talked about it, saying "Fred had a chassis with a completely new kind of suspension, and he took me for a ride in it. It was just a chassis with no body, and he had built it with his kids. I was so impressed with it that I told my boss at Ford . . . [Harley Copp, the chief engineer] about it, and arranged for Harley and his wife and my wife and me to go to Dayton, and Harley took a ride in it and was so impressed with it that he went back to Ford and persuaded them to buy it for its patent."

Fred led advanced engineering for the new, 1960 Ford Falcon, which was Ford's first "unibody" car. It had breakthrough design and sold incredibly well. (Photo courtesy Rex Gray, Creative Commons via Wikimedia Commons)

All the Ford engineers who worked on the Falcon team in Michigan. Fred is front row of four men, far left.

Ford Motor Company   FORD DIVISION

Intra-Company Communication   April 8, 1959

TO: Mr. Will Scott

cc: Messrs. J. D. Collins   R. L. Peters
                 H. F. Copp   D. E. Petersen
                 R. C. Graham   J. W. Richards
                 J. R. Hollowell   T. H. Risk
                 H. A. Matthias   J. P. Thornton
                 J. G. McQuaid

From: F. J. Hooven

Subject: Proposed Front-Engine Rear-Drive Vehicle

We have given very careful consideration to the letter received from
Mr. R. C. Versey in which he has proposed that the cost of our vehicles be
reduced by leaving the engine in its usual place at the front of the car,
but driving through the rear wheels instead of through the front wheels as
in conventional vehicles. Mr. Versey's proposal included building the final
drive and differential in unit with the rear axle and connecting this assembly to the power package by means of a long shaft passing from the front to
the rear of the vehicle under the floor. He furthermore proposed that this
rear axle assembly be completely unsprung and that the required motion relative to the vehicle be accommodated by universal joints at the extremities
of the driving shaft.

It is our conclusion that while Mr. Versey's proposed design exhibits a certain ingenuity, it is not feasible without intolerable compromises in package,
function and quality of the vehicle, and that certain proposed components
would present almost insurmountable engineering problems. It is also the
concurred opinion of Manufacturing, both at the Engine and Chassis Division
and the Ford Division that such a vehicle could not be manufactured in our
plants.

Some of the more detailed considerations are outlined below:

PACKAGE

Mr. Versey's sketch shows the transmission directly in the middle of the
front passenger compartment. Such a proposal is completely intolerable from
the package standpoint, but assuming that such a condition could not be avoided

Fred did not enjoy his time at Ford. This first page of a memo he wrote is trademark Fred. Accounting had sent a memo with suggestions for design changes that would reduce the cost of a vehicle design he had proposed. Fred replied with dripping sarcasm about their suggestions, which he deemed laughable. However, Accounting (whom Fred considered to be dimwits) took his memo seriously.

Fred in his study at his home in Vermont. He loved teaching at nearby Thayer School of Engineering at Dartmouth College.

ODYSSEY OF
A BORED
DIRECTOR

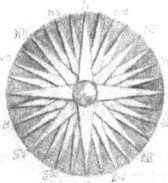

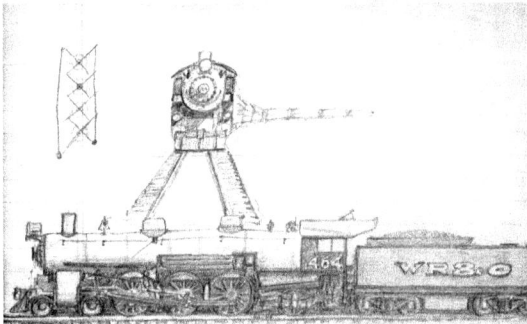

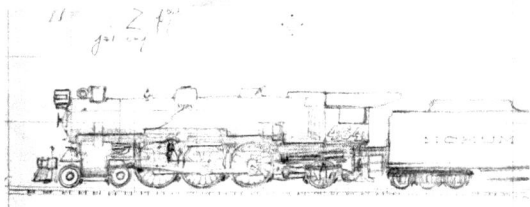

Fred served on the board of directors of the Yellow Springs Instrument Company. At meetings, he easily became bored and began to doodle. The directors loved his artwork and collected it into a booklet when he cycled off the board. Here are a few doodles from the booklet.

Reprinted from The Review of Scientific Instruments, Vol. 23, No. 12, 748-753, December, 1952
Printed in U. S. A.

A Large Capacity, All-Glass Dispersion Oxygenator and Pump*

Leland C. Clark, Jr., Frederick Hooven, and Frank Gollan†
Fels Research Institute for the Study of Human Development, Antioch College, Yellow Springs, Ohio
(Received May 9, 1951)

A large capacity all-Pyrex dispersion oxygenator and pump is described which is capable of substituting temporarily for the function of the heart and lungs. Blood is oxygenated by tiny bubbles of oxygen dispersed in the blood. These bubbles are coalesced on a polymethylsiloxane surface and eliminated in a bubble trap which also serves as the pumping chamber. Pumping is accomplished by an alternating suction and pressure controlled by two electrodes, suspended in the chamber, which actuate two solenoid valves. The apparatus causes no significant damage to the red blood cells. The apparatus is portable, easily cleaned and sterilized, and inexpensive. Oxygenation, carbon dioxide elimination, temperature, pulsation rate and pressure, flow, and mean pressure can be controlled for various physiological studies. The blood volume of the apparatus is constant except for the stroke volume and in this report resembles the normal heart.

SEVERAL methods for the oxygenation of blood have been reported and are currently being developed and tested in various laboratories throughout the world. Most of these methods[1-5] depend upon creating a large interface between oxygen and blood by transforming the blood into a thin moving film in an atmosphere of oxygen. The present oxygenator is based upon the knowledge that a huge gas-liquid interface can be created in a relatively small volume by bubbling gas through the liquid. Its inherent difficulty is that of completely removing these bubbles after their function is discharged. Although many methods, such as continuous centrifugation or even filtration, may be devised to remove these bubbles, the simpler method of coalescence and trapping is used in the present apparatus. The availability of a nontoxic, nearly water-insoluble, polymethylsiloxane "defoaming" compound has been utilized as the surface for this coalescence, but other compounds and other combinations of surface-active agents and defoaming substances may be used.

Previous work[6-8] has demonstrated the high efficiency of dispersion oxygenation as a means of sub-

stituting for the function of the lungs. The present report is confined to the description of an improved, simplified, all-Pyrex apparatus. Animal experiments are mentioned to illustrate particular points in the application of the instrument.

PRINCIPLE

The apparatus consists of an oxygen disperser, a coalescence chamber, and a pumping chamber, as shown in Figs. 1 and 2. Venous blood enters the bottom of the apparatus and flows upwards inside of a porous, fritted Pyrex tube where oxygen, in the form of very small bubbles, is mixed with it. The oxygen, under a pressure of 250 to 700 mm of mercury, is forced through the porous tube from the glass jacket which surrounds the tube. The blood is completely oxygenated by the time it leaves this chamber and needs only to be freed of gas before it can be returned to the animal. To accomplish this, the blood passes through a "coalescence chamber" filled with glass beads or polyethylene fiber thoroughly coated with a silicone defoaming compound. Here the small bubbles coalesce to form large bubbles. Since this is a surface phenomenon, all of the oxygen bubbles must contact some of the defoaming compound. When the chamber is properly designed, this is a remarkably rapid and efficient process. The coalesced bubbles, of course, contain not only excess oxygen but also the carbon dioxide released during the oxygenation process. These large bubbles and blood leave the coalescence unit and pass into the pumping chamber at the top where the bubbles emerge, leaving behind oxygenated, bubble-free blood to be returned to the animal.

The pumping is accomplished in the apparatus itself by applying alternating suction and pressure to the surface of the blood in the upper chamber. This pulsating action is controlled by two electrodes which are

* This work was supported by a grant from the American Heart Association.
† We wish to thank Mr. Richard Smith for his technical assistance.
[1] J. H. Gibbon, Jr., Arch. Surg. 34, 1105 (1937).
[2] T. C. Stokes and J. H. Gibbon, Jr., Surg. Gynecol. Obstet. 91, 138 (1950).
[3] V. O. Björk, Acta Chir. Scand. 96, Suppl. 137 (1948).
[4] J. Jongbloed, Nederland. Tijdschr. Geneesk. 92, 1065 (1948).
[5] Karlson, Dennis, Sanderson, and Culmer, Proc. Soc. Exp. Biol. Med. 71, 204 (1949).
[6] W. Kolff and C. F. Dubbelman, Geneesk. Gids, Van 30 (June 1949).
[7] Clark, Gollan and Gupta, Science 111, 85-87 (1950).
[8] Clark, Gupta, and Gollan, Proc. Soc. Exp. Bio. Med. 74, 268-271 (1950).
[9] Gollan, Clark, and Gupta, Am. J. Med. Sci. 222, 76-81 (1951).

748

Leland Clark invented the first commercial heart and lung machine, especially the vital oxygen sensor that rendered it a breakthrough machine. Fred played a pivotal role in the machine and its success— it was he who engineered, designed, and built it, even if he did not invent the breakthrough componentry. Here is the paper he co-authored with Clark.

The Harris-Intertype Phototypesetter ("Fotosetter") came out in the late 1950s. A website dedicated to printing technology history says, "It was the first photo typesetting machine and was based upon the standard intertype machine, replacing the brass type matrices with small film negatives and instead of casting, used these to expose photographic paper." For Harris-Intertype, Fred invented the digital phototypesetter, which was a bestseller for years.

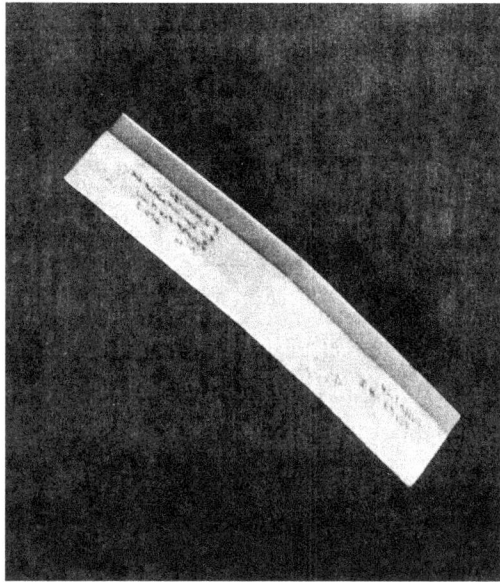

Fred won the "endurance" category—nearly 11,000 entries—for Scientific American's *first ever paper airplane contest. He based the design of his plane on what he learned from his friends, the Wright Brothers. As is evident from this image, it was laughably simple; it didn't even look like a plane.*

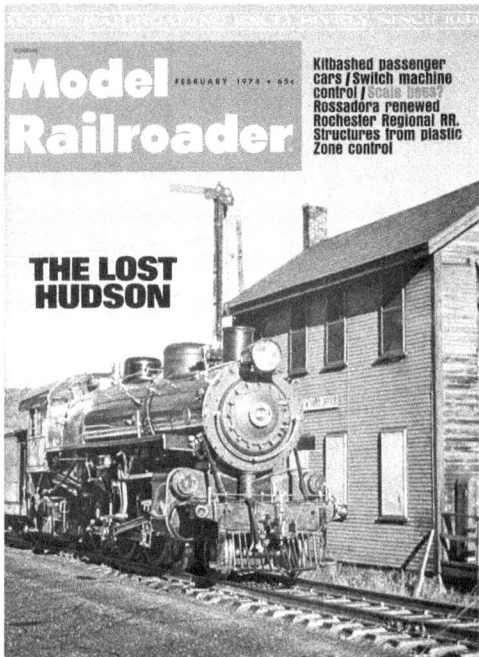

Fred built a model train and took a photograph of it at the Rutland, Vermont train station. To achieve the perspective and dimensions that would make the train look like an actual, life-sized train, Fred did various calculations and rebuilt the lens of the camera he used. Readers who did not know otherwise would think the train was a real, old-fashioned train at the modern-day train station.

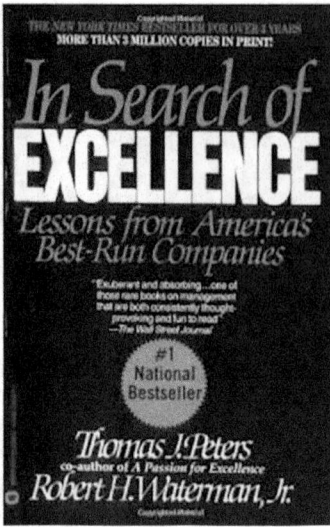

subject, material, or process is unmatchable in the abstract, via paper analysis or description.

Thus, when "touch it," "taste it," "smell it" become the watchwords, the results are most often extraordinary. Equally extraordinary are the lengths to which people will go to avoid the test-it experience. Fred Hooven, protégé of Orville Wright, holder of thirty-eight major patents, and senior engineering faculty member at Dartmouth, describes a ludicrous, yet all-too-typical, case: "I can think of three instances in my career in which my client was making no progress on a complicated mechanical problem, and I insisted that the engineers and the technicians [model builders] be put in the same room. In each case the solution came rapidly. *One objection I remember being offered was that if we put the engineers in the same room with the shop it would get the drawings dirty.*" Hooven adds, in support of the overall point, "The engineer must have immediate and informal access to whatever facilities he needs to put his ideas into practice. . . . It costs more to make drawings of a piece than to make the piece, and the drawing is only one-way communication, so that when the engineer gets his piece back he has probably forgotten why he wanted it, and will find out that it doesn't work because he made a mistake in the drawings, or that it needs a small change in some respect, which too often takes another four months to make right."

So, via experimentation, it is much easier for people (e.g., designers, marketers, presidents, salesmen, customers) to think creatively about a product, or be creative about product uses, if a prototype, which is to say a low level of abstraction, is in hand. Thus, no

Author and business guru Tom Peters wrote about Fred's inventive mindset in his mega-bestseller, In Search of Excellence.

Fred helped invent and engineer the Dartmouth Digital Synthesizer, which was commercialized as the Synclavier, a breakthrough digital music synthesizer used by some of the top musicians and movie-score producers ever. It hit many firsts— the first digital synthesizer you could buy, the first commercial synthesizer to use purely digital (not analog) sound generation, and also the world's first commercial FM synthesizer. Today, many music enthusiasts consider it the "Mercedes" of the category.

Acknowledgement

This account owes its entire existence to Frederick Allan Goerner, author of *The Search for Amelia Earhart*. The documentation on which it is based constitutes the most complete collection in existence on the subject of Miss Earhart's last flight and the subsequent search for her and Fred Noonan, her navigator, and it was all collected by Goerner. There are official reports and documents, letters, interviews, questionnaires, radio logs, photographs, maps and techcial /sic/ memoranda, he has them all.

Gathered largely since the publication of his book in 1966, this material has been made available to me in its entirety. While the accompanying conclusions are entirely mine, and not necessarily shared by Goerner, they owe much to the lively discussion that has gone on between us since our acquaintance began 15 years ago. It has been a source of abiding interest and satisfaction for me, and I am deeply grateful to him for having made it all possible.

Frederick J. Hooven
Norwich, Vermont
June, 1982

A Personal Note

Before Miss Earhart took off on her Round-the-World flight she removed from her plane a modern radio compass that had been installed and replaced it with an older, lighter-weight model of much less capability. I am the engineer who had invented and developed the radio compass that was removed, and I discussed its features with Miss Earhart before the installation was made. I have reason to believe that it was the failure of her radio direction-finder to do what the more modern model could have done that caused her to be lost. The story is told herein, and it is plain to see why I have been so very much interested in the subject.

I met Miss Earhart for lunch at Wright Field in the summer of 1936. She was accompanied by a younger woman flyer, quite unknown at the time, Jacqueline Cochran. Although she moved in a man's world, and wore men's trousers and wore a short haircut, there was nothing masculine about Miss Earhart. Every inch a lady, she was gracious and quiet-spoken, thoroughly feminine and attractive.

Too much time has elapsed for me to remember when it was that I learned that my device was not on the Earhart plane when it was lost, or even whether it was before or after the takeoff that I learned. But I have been possessed by the desire to know what did happen, and by the wish that things had happened differently.

Amelia Earhart's Last Flight

In the course of her West-to-East flight around the world, Amelia Earhart and her navigator, Fred Noonan, departed Lae, New Guinea, in her Lockheed Electra "Flying Laboratory" bound for Howland Island, a tiny dot 2626 miles East by North. Awaiting her there was the U. S. Coast Guard Cutter *Itasca* with a crew to service her plane and instructions to provide radio guidance and communication, for Howland was too small a target for a navigator to hit without the aid of a radio direction finder tuned to a guidance station at the island. The *Itasca* was equipped with such a station and the plane carried a direction-finder on board. It was July 2, 1937.

Howland wasn't the ideal spot for a round-the-world flight, but it was U.S. territory and it had a runway. At that time Japan had a mandate from the League of Nations for the Marshalls, the Mariannas and the Caroline Islands and had lately closed all this territory to the rest of the world, being known to be fortifying them in preparation for future military actions, as events later proved. The military were firmly in power in Japan then, and they were alarmed by the news of the projected Earhart flight, suspecting every move as a possible effort at espionage. As a result, Japanese "fishing boats" were positioned along the course, observing the flight. The United States, in the grip of a deep depression and growing isolationism, had no stomach for any warlike moves, and save for the *Itasca*, the ancient coalburning tug *Ontario* stationed between Lae and Howland and the mine sweeper *Swan* between Howland and Hawaii, no U.S. ship was nearer than Hawaii.

Fred wrote The Hooven Report in 1982. This is an excerpt. In the report, he recounted meeting Amelia Earhart, and he performed very good calculations regarding where she might have landed or crashed. He nonetheless came to a controversial and (by his own later admission) wrong conclusion. Still, he enjoyed researching the Earhart disappearance mystery.

Notes

Interviews by author with others are largely not cited here but are covered in Sources, below.

<u>Young Man</u>
"The most indifferent . . ." comes from Cassidy Boyer, "Why Is Dayton Called the 'Gem City?'" *Dayton-Daily-News*, Sept. 23, 2006.

"I could get . . ." comes from Fred Hooven, "Reminiscences of an Old Hi-Fi Hound," Private Papers, August, 1975.

"In 1922 I . . ." comes from "Reminiscences of an Old Hi-Fi Hound."

<u>Growing Family</u>
"For my part . . ." comes from Fred Hooven, "Penetrating Analysis Made for Times Study," *Kettering-Oakwood Times*, January 17, 1957.

<u>Inventive Engineer</u>
"In 1940 I . . ." comes from "Reminiscences of an Old Hi-Fi Hound."

<u>Independent Inventor</u>
"Overzealous business practices . . ." comes from Cameron Warren. "Fred Hooven: The Man Who (Nearly) Found Earhart." Amelia Earhart Society Newsletter, 1997. Republished on the Earharttruth blog, Feb 1, 2019. https://earharttruth.wordpress.com/2019/02/01/fred-hooven-man-who-nearly-found-earhart/.

"The first blind . . ." comes from Fred Hooven, Letter to Myron Tribus, Sept. 30, 1966, Fred Hooven Papers, Rauner Special Collections Archives, Dartmouth College.

"The actual performance . . ." comes from "The Norden Bombsight: Was It Truly Accurate Beyond Belief?" Warfare History Network, 21 July 2016.

"Hooven said first . . ." comes from Jack Jones, "First Bombing Computer Goes to AF Museum," *Dayton Daily News,* Dec. 18, 1970.

"In September 1944 . . ." comes from H. G. Sennert, "History of Shoran," *Programmatic Engineering and Remote Sensing*, Imaging and Geospatial Information Society, date unknown.

"Inventor. Development of . . ." comes from Frederick J. Hooven,

"Résumé," Private Papers, no date.

A bureau of . . ." comes from Ric Gillespie. Comment in Amelia Earhart forum. International Group for Historic Aircraft Recovery (TIGHAR) website. Feb 28, 2009.

"Hooven later alleged . . ." ibid.

"Hooven's DF (direction finder) . . ." comes from "Fred Hooven: The Man Who (Nearly) Found Earhart."

"It looks, to me, . . ." comes from biography page of callsign K4RNN,

"Ray." Amateur radio callsign website QRZ. No date.

"In the report . . ." comes from Frederick Hooven, "Amelia Earhart's Last Flight," aka "The Hooven Report," 1982, archived at the Smithsonian National Air & Space Museum Library and reproduced on the TIGHAR website. https://tighar.org/Projects/Earhart/Archives/Documents/Hooven_Report/HoovenReport.html.

"Despite a minor flaw . . ." comes from Cameron Warren, "Fred Hooven: The Man Who (Nearly) Found Earhart."

"The idea of the Japanese mounting . . ." comes from William Trail, comment on "Fred Hooven: The Man Who (Nearly) Found Earhart," republished Feb. 1, 2019, on Earharttruth.wordpress.com.

"Photographic Type Composition" comes from US patent 2966835A Hooven, Frederick J., and Richard C. O'brien. Photographic type composition, filed May 27, 1957, issued January 3, 1961.

"Blood is oxygenated . . ." comes from Clark, Leland C., et al.,

"A Large Capacity, All-Glass Dispersion Oxygenator and Pump." *Review of Scientific Instruments, vol. 23, no. 12*, American Institute of Physics, Dec. 1952, pp. 748-53.

Iconic Advisor
"Fred Hooven was . . ." comes from a published magazine interview, Hooven Papers. Magazine unknown. Date unknown.

"We have given . . ." and "In conclusion I . . ." come from Frederick Hooven, "Proposed Front-Engine Rear-Drive Vehicle, April 8, 1959," Intra-Company Communication/Ford Motor Company, Private Papers.

"The thing I remember most . . ." comes from Myron Tribus, "Frederick Johnson Hooven, 1905-1985," *NAE Memorial Tributes, Volume 3*, 1989.

"Fred Hooven was . . ." comes from a published magazine interview, Hooven Papers. Magazine unknown. Date unknown.

"Mr. Wright loved . . ." comes from "Crafty Ford Official Wins World's Paper Plane Title." *Detroit Free Press*, 23 Feb. 1967.

"Dear Fred, Your . . ." comes from "Odyssey of a Bored Director," Yellow Springs Instrument Company, Private Papers, no date.

Beloved Professor
"When touch it . . ." comes from Tom Peters and Robert H. Waterman, *In Search of Excellence: Lessons from America's Best-Run Companies*, Harper Business, 1982.

"Of all the people . . ." and quotes throughout the paragraph come from Interview with John Collier, video (Zoom), Sept. 18, 2020.

"Wilbur and Orville . . ." comes from Frederick J. Hooven, "The Wright Brothers' Flight-Control System," *Scientific American*, November 1, 1978.

"It did so . . ." comes from "Thayer Professor and Student Invent the Synclavier." 1975 page on Dartmouth College 250th anniversary website. No date. https://250.dartmouth.edu/highlights/thayer-professor-and-student-invent-synclavier

"The LinkedIn Learning . . ." comes from "Learning the Arturia Collection—Synclavier V History." LinkedIn Learning. No date. https://www.linkedin.com/learning/learning-the-arturia-collection/synclavier-v-history

"The invention provides" comes from Hooven, Frederick J., and Robert B. Lowd. Optical fiber coupling system. European Union EP0010388A1, filed October 5, 1979, and issued April 30, 1980. https://patents.google.com/patent/EP0010388A1/en?inventor=Frederick+J.+Hooven&oq=Frederick+J.+Hooven&page=4.

Grandfather
Perhaps the first . . ." comes from Myron Tribus, "Frederick Johnson Hooven, 1905-1985."

Legacy
"Mr. Hooven's inventions . . ." comes from Newspapers.com.

"8 Feb 1985, 21 - The Boston Globe at Newspapers.Com." Accessed October 27, 2020. http://www.newspapers.com/image/437501549/?terms=Frederick%20Hooven&match=1.

"Fred Hooven enjoyed . . ." ibid.

Sources

I referred to many materials to learn and write about Fred Hooven. These include:

Recorded and transcribed interviews with family and friends, namely Michael Hooven, Mike Hooven, Martha (Hooven) Richardson, and John Collier.

Frederick J. Hooven papers, privately owned.

Rauner Special Collections Library, Archives at Dartmouth College.

Many websites, newspapers, magazines, books and more, all digital, some listed in Notes; more complete list available upon request.

Acknowledgments

I'd like to thank Mike Hooven, who compiled much of the information for this short book and who served as development and content editor. Also, his wife, Sue, provided valuable feedback. Mike's vision for the book, which we dubbed a "booklet" while we worked together, and his eagerness to see it completed in timely fashion are wonderful gifts in the world of collaboration. I also thank the friends, colleagues, and family members of Fred Hooven for speaking with me about their experiences with him. These include Mike Hooven, Sr., Martha (Hooven) Richardson, Steve Richardson, and John Collier. The reference librarians, Morgan and Scout, at the Rauner Special Collections Library at Dartmouth College made quick and easy work of accessing archived materials. Copy editor Patricia Fernberg, book designer Ben Small, and publishing manager David Gray are an easy and seasoned team to work with; my thanks to them. Lastly, I thank you, beloved reader.

www.ingramcontent.com/pod-product-compliance
Lightning Source LLC
LaVergne TN
LVHW051421080426
835508LV00022B/3180